LEARN
Adobe Premiere Pro CC for Video Communication

SECOND EDITION

Adobe Certified Associate Exam Preparation

Joe Dockery
Condrad Chavez
with Rob Schwartz

LEARN ADOBE PREMIERE PRO CC FOR VIDEO COMMUNICATION, SECOND EDITION
ADOBE CERTIFIED ASSOCIATE EXAM PREPARATION

Copyright © 2019 by Pearson Education, Inc. or its affiliates. All Rights Reserved.

Adobe Press is an imprint of Pearson Education, Inc.
For the latest on Adobe Press books and videos, go to www.adobepress.com.
To report errors, please send a note to errata@peachpit.com. For information regarding permissions, request forms and the appropriate contacts within the Pearson Education Global Rights & Permissions department, please visit www.pearsoned.com/permissions/.

Adobe Press Editor: Laura Norman
Development Editor: Steve Nathans-Kelly
Senior Production Editor: Tracey Croom
Compositor: Kim Scott, Bumpy Design
Copyeditor: Liz Welch
Proofreader: Kim Wimpsett
Indexer: James Minkin
Cover & Interior Design: Mimi Heft
Cover Illustration: Windesign/ShutterStock

NOTICE OF RIGHTS
If this guide is distributed with software that includes an end user license agreement, this guide, as well as the software described in it, is furnished under license and may be used or copied only in accordance with the terms of such license. Except as permitted by any such license, no part of this guide may be reproduced, stored in a retrieval system, or transmitted, in any form or by any means, electronic, mechanical, recording, or otherwise, without the prior written permission of Adobe Systems Incorporated. Please note that the content in this guide is protected under copyright law even if it is not distributed with software that includes an end user license agreement.

NOTICE OF LIABILITY
The content of this guide is furnished for informational use only, is subject to change without notice, and should not be construed as a commitment by Adobe Systems Incorporated. Adobe Systems Incorporated assumes no responsibility or liability for any errors or inaccuracies that may appear in the informational content contained in this guide.

Please remember that existing artwork or images that you may want to include in your project may be protected under copyright law. The unauthorized incorporation of such material into your new work could be a violation of the rights of the copyright owner. Please be sure to obtain any permission required from the copyright owner. Any references to company names in sample files are for demonstration purposes only and are not intended to refer to any actual organization.

TRADEMARKS
Adobe, the Adobe logo, Adobe Certified Associate, Creative Cloud, the Creative Cloud logo, Adobe Premiere Pro, Animate, Flash, Illustrator, Photoshop, Dreamweaver, Adobe Capture, Flash Player, and Typekit are registered trademarks of Adobe Systems Incorporated in the United States and/or other countries. All other trademarks are the property of their respective owners.

Apple, Mac OS, macOS, and Macintosh are trademarks of Apple, registered in the U.S. and other countries. Microsoft and Windows are either registered trademarks or trademarks of Microsoft Corporation in the U.S. and/or other countries.

Unless otherwise indicated herein, any third-party trademarks that may appear in this work are the property of their respective owners and any references to third-party trademarks, logos, or other trade dress are for demonstrative or descriptive purposes only. Such references are not intended to imply any sponsorship, endorsement, authorization, or promotion of Pearson Education, Inc. products by the owners of such marks, or any relationship between the owner and Pearson Education, Inc. or its affiliates, authors, licensees or distributors.

ISBN-13: 978-0-13-487857-7
ISBN–10:   0-13-487857-4

I would like to dedicate this book to my father, Robert Patterson, who was always there for me. His example of hard work and dedication to education has been a guiding light in my life.

—Joe Dockery

To Sarah, who makes everything better.

—Conrad Chavez

Acknowledgments

I wish to personally thank the following people for their contributions to creating this book:

My loving and understanding wife, Laura, who puts up with all the crazy projects I get myself into. Thank you for your patience and support. My family and students (Truman, Mason, Colten, and Liam) for helping me create all the assets for the book. My good friend, and founder of Brain Buffet, Rob Schwartz, who has been the driving force behind this book series. Thank you for your guidance and encouragement throughout the authoring process. My editor, Steve Nathans-Kelly, thanks for patiently catching all my errors. Your input made the book stronger. Lisa Deakes and the entire Adobe Education Leader crew for all your support. The Snoqualmie Valley School District and all my amazing students over the years. You have shaped the teacher and author that I am today.

—Joe Dockery

I'd like to thank Laura Norman at Peachpit for her support and encouragement, and Steve Nathans-Kelly for the editorial guidance that made this project go smoothly.

—Conrad Chavez

About the Authors

Joe Dockery (video author) has taught for 25 years in the Snoqualmie Valley School District and currently leads the Digital Media Academy at Mount Si High School. He engages his students in real-world design projects from their school and community to ensure they receive authentic learning experiences. As an Adobe Education Leader, Joe Dockery consults and trains nationwide on the use and integration of Adobe software. His awards include the Washington State Golden Apple Award, Radio Shack National Technology Teacher of the Year Award, Educator of the Year Award from the Snoqualmie Valley Schools Foundation, International Society for Technology in Education (ISTE)'s "Best of the Best" and "Making IT Happen" awards, Adobe Education Leader "Impact" Award, and Give Good Awards: Educational Excellence 2015. Joe is an Adobe Certified Associate in Premiere Pro CC.

Conrad Chavez (book author) is an author and photographer with over two decades of experience with Adobe digital media workflows. During his time at Adobe Systems Inc., Conrad helped write the user guide for Adobe Premiere (the precursor to Adobe Premiere Pro). He is the author of several titles in the Real World Adobe Photoshop and Adobe Classroom in a Book series, and he writes articles for websites such as CreativePro.com and peachpit.com. Visit his website at conradchavez.com.

Rob Schwartz, author of *Learn Adobe Photoshop CC for Visual Design* and contributor to the Learn series, is an award-winning teacher (currently at Sheridan Technical College in Hollywood, FL) with more than 15 years' experience in technical education. Rob holds several Adobe Certified Associate certifications and is also an Adobe Certified Instructor. As an Adobe Education Leader Rob won the prestigious Impact Award from Adobe, and in 2010 Rob was the first Worldwide winner of the Certiport Adobe Certified Associate Championship. Find out more about Rob at his online curriculum website at brainbuffet.com.

Contents

Getting Started viii

1 Introduction to Adobe Premiere Pro CC 3

About Adobe Learn Books 4
Managing Files for Video Production 5
Downloading, Unpacking, and Organizing 10
Identifying Job Requirements 13
Starting Premiere Pro 15
Setting Up the New Project Dialog Box 18
Locating a Project and Editing Its Settings 24
Exploring Panels and Workspaces 25
Using Workspaces 35
Using Premiere Pro on a PC or Mac 36
Importing Media 38
Understanding a Basic Editing Workflow 43
Editing a Sequence 44
Navigating the Timeline 56
Exploring the Editing Tools 57
Working with Audio 66
Adding a Simple Title 70
Using Video Transitions and Effects 75
Exporting a Finished Video File 80
Challenge 87
Conclusion 87

2 Editing an Interview 89

Preproduction 89
Setting Up the Interview Project 90
Creating the Interview Sequence 95
Diving Deeper into the Workspace 99
Making Quick Fixes to Audio 105
Making Quick Fixes to Color 110
Subtracting Unwanted Clip Segments 114
Getting Organized in the Timeline Panel 119
Applying L and J Cuts 120
Playing a Clip Faster or Slower 123
Playing a Sequence Smoothly 125
Varying Clip Speed Over Time 127
Using Markers 131
Adding Titles 133
Stabilizing a Shaky Clip 139
Merging Separate Video and Audio Files 142
Exporting with Adobe Media Encoder CC 143
Challenge: Mini-Documentary 147

3 Editing an Action Scene 149

Getting Ready in Preproduction 149
Acquiring and Creating Media 151
Taking Another Look at Preferences 154
Setting Up the Action Scene Project 158
Importing Files and Maintaining Links 159
Inspecting the Properties of a Clip 163
Starting a Rough Cut 164
Editing with Vertical Video 168
Editing a Multicam Sequence 172
Finishing Sequence Edits 176
Sweetening Different Audio Types 177
Applying an Adjustment Layer 180
Reviewing Timeline Controls 182
Adding Credits 183
Recording a Voiceover 187
Nesting Sequences for Different Delivery Requirements 190

Using Proxies and Removing Unused Clips   193
Exporting Multiple Sequences   197
Using the Project Manager   199
Challenge   201

4 Compositing with Green Screen Effects   203

Preproduction   203
Setting Up a Project   204
Compositing a Green Screen Clip with a New Background   207
Adding and Animating More Graphics   212
Exporting Final Video and Audio   219
Challenge: Create Your Own Composited Video   221
Conclusion   221

5 Creating a Video Slide Show   223

Preproduction   223
Setting Up a Slide Show Project   224
Creating a Sequence from Multiple Files Quickly   225
Adding a Ken Burns Motion Effect   230
Exporting Multiple Versions with Adobe Media Encoder   232
Challenge: Your Own Slide Show   236
Conclusion   236

6 Working in the Video Industry   239

Phases of Production   239
Reviewing Job Requirements   243
Roles of a Video Production Team   246
Communicating Effectively   248
Visual Standards and Techniques   250
Licensing, Rights, and Releases   263
Moving Into the Industry   266

7 Wrapping It Up!   269

Extending Premiere Pro CC with Adobe Creative Cloud   269
Where to Go Next   277
Good Luck, and Have Fun!   278

ACA Objectives Covered   279
Glossary   284
Index   289

Getting Started

Welcome to *Learn Adobe Premiere Pro CC for Video Communication*! We use a combination of text and video to help you learn the basics of video editing with Adobe Premiere Pro CC along with other skills that you will need to get your first job as a video editor. Adobe Premiere Pro CC is a powerful program for capturing footage from a variety of devices and assembling it into professional-quality video with sophisticated transitions, special effects, and text. You can also use Premiere Pro to export your video to many popular formats that your viewers can watch on a wide range of screens, including desktop computers and mobile devices like phones and tablets.

About this product

Learn Adobe Premiere Pro CC for Video Communication was created by a team of expert instructors, writers, and editors with years of experience in helping beginning learners get their start with the cool creative tools from Adobe Systems. Our aim is not only to teach you the basics of the art of video editing with Premiere Pro but to give you an introduction to the associated skills (like design principles and project management) that you'll need for your first job.

We've built the training around the objectives for the Video Communication Using Adobe Premiere Pro CC (2018) Adobe Certified Associate Exam. If you master the topics covered in this book and video, you'll be in good shape to take the exam. But even if certification isn't your goal, you'll still find this training will give you an excellent foundation for your future work in video. To that end, we've structured the material in the order that makes most sense for beginning learners (as determined by experienced classroom teachers), rather than following the more arbitrary grouping of topics in the ACA Objectives.

To aid you in your quest, we've created a unique learning system that uses video and text in partnership. You'll experience this partnership in action in the Web Edition, which lives on your Account page at peachpit.com. The Web Edition contains 8 hours of video—the heart of the training—embedded in an online eBook that supports the video training and provides background material. The eBook material is also available separately for offline reading as a printed book or an eBook in a variety of formats. The Web Edition also includes hundreds of interactive review questions you can use to evaluate your progress. Purchase of the book in *any* format entitles you to free access to the Web Edition (instructions for accessing it follow later in this section).

Most chapters provide step-by-step instructions for creating a specific project or learning a specific technique. Other chapters acquaint you with other skills and concepts that you'll come to depend on as you use the software in your everyday work. Many chapters include several optional tasks that let you further explore the features you've already learned.

Each chapter opens with two lists of objectives. One list lays out the learning objectives: the specific tasks you'll learn in the chapter. The second list shows the ACA exam objectives that are covered in the chapter. A table at the end of the book guides you to coverage of all of the exam objectives in the book or video.

Most chapters provide step-by-step instructions for creating a specific project or learning a specific technique. Many chapters include several optional tasks that let you further explore the features you've already learned. Chapter 6 acquaints you with other skills and concepts that you'll come to depend on as you use the software in your everyday work. Here is where you'll find coverage of parts of Domain 1 of the ACA Objectives that don't specifically relate to features of Premiere Pro but that are important components of the complete skill set that the ACA exam seeks to evaluate.

Conventions used in this book

This book uses several elements styled in ways to help you as you work through the exercises.

Text that you should enter appears in bold, such as:

In the Link field in the Property inspector, type **https://helpx.adobe.com/premiere-pro.html**.

Terms that are defined in the Glossary appear in bold and in color, such as:

> The **web font** that's used in the header of the page is just what the client is looking for. That's a great thing.

▶ *Video 4.6* Create Picture-in-Picture

Links to videos that cover the topics in depth appear in the margins.

The ACA Objectives covered in the chapters are called out in the margins beside the sections that address them.

★ *ACA Objective 2.4*

Notes give additional information about a topic. The information they contain is not essential to accomplishing a task but provides a more in-depth understanding of the topic.

> **NOTE**
> *A histogram is a graph that represents how many pixels of each tonal value exist within the image.*

Operating system differences

In most cases, Premiere Pro CC works the same in both Windows and macOS. Minor differences exist between the two versions, mostly due to platform-specific issues. Most of these are simply differences in keyboard shortcuts, how dialogs are displayed, and how buttons are named. In most cases, screen shots were made in the macOS version of Premiere Pro and may appear somewhat differently from your own screen.

Where specific commands differ, they are noted within the text. Windows commands are listed first, followed by the macOS equivalent, such as Ctrl+C/Cmd+C. In general, the Windows Ctrl key is equivalent to the Command (or Cmd) key in macOS and the Windows Alt key is equivalent to the Option (or Opt) key in macOS.

As lessons proceed, instructions may be truncated or shortened to save space, with the assumption that you picked up the essential concepts earlier in the lesson. For example, at the beginning of a lesson you may be instructed to "press Ctrl+C/Cmd+C." Later, you may be told to "copy" text or a code element. These should be considered identical instructions.

If you find you have difficulties in any particular task, review earlier steps or exercises in that lesson. In some cases, if an exercise is based on concepts covered earlier, you will be referred to the specific lesson.

Installing the software

Before you begin using *Learn Adobe Premiere Pro CC for Video Communication,* make sure that your system is set up correctly and that you've installed the proper software and hardware. This material is based on the original 2018 release of Adobe Premiere Pro CC (version 12.1) and is designed to cover the objectives of the Adobe Certified Associate Exam for that version of the software.

The Adobe Premiere Pro CC software is not included with this book; it is available only with an Adobe Creative Cloud membership, which you must purchase or which must be supplied by your school or other organization. In addition to Adobe Premiere Pro CC, some lessons in this book have steps that can be performed with Adobe Media Encoder and other Adobe applications. You must install these applications from Adobe Creative Cloud onto your computer. Follow the instructions provided at helpx.adobe.com/creative-cloud/help/download-install-app.html.

ADOBE CREATIVE CLOUD DESKTOP APP

In addition to Adobe Premiere Pro CC, this training requires the Adobe Creative Cloud desktop application, which provides a central location for managing the many apps and services that are included in a Creative Cloud membership. You can use the Creative Cloud desktop application to sync and share files, manage fonts, access libraries of stock photography and design assets, and showcase and discover creative work in the design community.

The Creative Cloud desktop application is installed automatically when you download your first Creative Cloud product. If you have Adobe Application Manager installed, it auto-updates to the Creative Cloud desktop application.

If the Creative Cloud desktop application is not installed on your computer, you can download it from the Download Creative Cloud page on the Adobe website (creative.adobe.com/products/creative-cloud) or the Adobe Creative Cloud desktop apps page (www.adobe.com/creativecloud/catalog/desktop.html). If you are using software on classroom machines, be sure to check with your instructor before making any changes to the installed software or system configuration.

CHECKING FOR UPDATES

Adobe periodically provides updates to software. You can easily obtain these updates through the Creative Cloud. If these updates include new features that affect the content of this training or the objectives of the ACA exam in any way, we will post updated material to peachpit.com.

Accessing the free Web Edition and lesson files

Your purchase of this product in any format includes access to the corresponding Web Edition hosted on peachpit.com. The Web Edition contains the complete text of the book augmented with hours of video and interactive quizzes.

To work through the projects in this product, you will first need to download the lesson files from peachpit.com. You can download the files for individual lessons or download them all in a single file.

If you purchased an eBook from peachpit.com or adobepress.com, the Web Edition will automatically appear on the Digital Purchases tab on your Account page. Continue reading to learn how to register your product to get access to the lesson files.

If you purchased an eBook from a different vendor or you bought a print book, you must register your purchase on peachpit.com:

1. Go to www.peachpit.com/register.
2. Sign in or create a new account.
3. Enter ISBN **9780134878577**.
4. Answer the questions as proof of purchase.
5. The **Web Edition** will appear under the Digital Purchases tab on your Account page. Click the Launch link to access the product.

The **Lesson Files** can be accessed through the Registered Products tab on your Account page. Click the Access Bonus Content link below the title of your product to proceed to the download page. Click the lesson file links to download them to your computer.

Project fonts

All fonts used in these projects either are part of standard system installations or can be downloaded from Typekit, an Adobe service that is included with your Creative Cloud membership. (Typekit might not be included with some educational or institutional Creative Cloud memberships.)

Additional resources

Learn Adobe Premiere Pro CC for Video Communication is not meant to replace documentation that comes with the program or to be a comprehensive reference for every feature. For comprehensive information about program features and tutorials, refer to these resources:

- **Adobe Premiere Pro Learn & Support:** helpx.adobe.com/support/premiere-pro.html is where you can find and browse Help and Support content on Adobe.com. You can go there from inside Adobe Premiere Pro by choosing Help > Adobe Premiere Pro Help. Help is also available as a printable PDF document. Download the document at helpx.adobe.com/pdf/premiere_pro_reference.pdf.

- **Adobe Premiere Pro Tutorials:** helpx.adobe.com/premiere-pro/tutorials.html is where you learn more about Premiere Pro by watching many online video tutorials. You can go there from inside Premiere Pro by choosing Help > Adobe Premiere Pro Tutorials.

- **Adobe Premiere Pro In-App Tutorials:** There are a number of interactive tutorials built into Premiere Pro, which lead you through basic concepts using the software itself. To try them, inside Premiere Pro choose Help > Help > Adobe Premiere Pro In-App Tutorials.

- **Adobe Forums**: forums.adobe.com/community/premiere lets you tap into peer-to-peer discussions, questions, and answers on Adobe products.
- **Adobe Premiere Pro CC product home page**: adobe.com/products/premiere provides information about the current version's new features. It also provides links to related Adobe applications and to the Learn & Support page.
- **Adobe Add-ons**: adobeexchange.com/creativecloud.html is a central resource for finding tools, services, extensions, code samples, and more to supplement and extend your Adobe products.
- **Resources for educators**: adobe.com/education and edex.adobe.com offer a treasure trove of information for instructors who teach classes on Adobe software at all levels.

Adobe certification

The Adobe training and certification programs are designed to help video editors, designers, and other creative professionals improve and promote their product-proficiency skills. The Adobe Certified Associate (ACA) is an industry-recognized credential that demonstrates proficiency in Adobe digital skills. Whether you're just starting out in your career, looking to switch jobs, or interested in preparing students for success in the job market, the Adobe Certified Associate program is for you! For more information visit edex.adobe.com/aca.

Resetting preferences to their default settings

Premiere Pro lets you determine how the program looks and behaves (like tool settings and the default unit of measurement) using the extensive options in Edit > Preferences (Windows) or Premiere Pro CC > Preferences (macOS). To ensure that the preferences and default settings of your Adobe Premiere Pro program match those used in this book, you can reset your preference settings to their defaults. If you are using software installed on computers in a classroom, don't make any changes to the system configuration without first checking with your instructor.

To reset your preferences to their default settings, follow these steps:

1 Quit Adobe Premiere Pro.
2 Hold down the Alt key (Windows) or Option key (macOS).
3 Continue to hold the key and start Adobe Premiere Pro CC.
4 When the program's splash screen appears, release the key.

CHAPTER OBJECTIVES

Chapter Learning Objectives

- Set up and manage project files.
- Open and save Premiere Pro projects.
- Learn scratch disk options.
- Explore the Premiere Pro user interface.
- Learn basic panel functions.
- Customize your workspace.
- Manage files in the Project panel.
- Use the Source, Program, and Timeline panels to edit a sequence.
- Navigate the Timeline panel.
- Learn the functions of Premiere Pro tools.
- Add audio to a sequence.
- Add a title to a sequence.
- Export a sequence.

Chapter ACA Objectives

For full descriptions of the objectives, see the table on pages 279–283.

DOMAIN 1.0
WORKING IN THE VIDEO INDUSTRY
1.1, 1.1a, 1.2, 1.3, 1.4

DOMAIN 2.0
PROJECT SETUP AND INTERFACE
2.1, 2.1a, 2.1b, 2.2, 2.2a, 2.2b, 2.2c,
2.3, 2.3a, 2.3c, 2.4, 2.4a, 2.4b

DOMAIN 3.0
ORGANIZATION OF VIDEO PROJECTS
3.1, 3.1a, 3.2, 3.2a

DOMAIN 4.0
CREATE AND MODIFY VISUAL ELEMENTS
4.1, 4.1a, 4.2, 4.2a, 4.2b, 4.2c, 4.3, 4.3a, 4.3b

CHAPTER 1

Introduction to Adobe Premiere Pro CC

On the surface, video editing might seem to be about mastering a video editing application. But successful video editing—even with an application as powerful and versatile as Adobe Premiere Pro CC—is often about much more than just pushing the right buttons on the computer. Video production typically involves a high degree of both integration and collaboration.

Integration means creating a seamless video **program** by pulling together media from multiple sources, such as conventional video cameras, smartphones, drones, action cameras, microphones, stock footage, music, graphics, and still images.

▶ *Video 1.1 Welcome to the Team*

Collaboration is often required because the many elements that go into a **project** are typically created by a wide range of specialists—such as camera operators and audio recording engineers—and are coordinated by a producer. You will work with them as a team, so working successfully includes coordinating and cooperating with everyone on the team. That requires clear communication about standards and procedures.

In the next chapter, you'll complete a project in which you'll act as a member of the Brain Buffet production team, working with them to create a 15-second promotional video for a client's online newsletter (**Figure 1.1**). In this chapter, you'll organize the media assets you'll use in the newsletter project. In the process, you'll get an introduction to the Premiere Pro user interface and some of the things you can accomplish with it.

Figure 1.1 Working on the promo project

About Adobe Learn Books

Let's take just a second to explain what I'm trying to accomplish so we can be sure we're on the same page (pun intended!). Here's what I (and the other authors) hope to accomplish in this series.

Have fun

This is an important goal for me, as I hope it is for you! When you're having fun, you learn more, and you're more likely to remember what you're learning. Having fun also makes it easier to focus and stick with the task at hand.

Even if the projects you create as you complete the exercises in this book aren't the kinds of things you'd create on your own, I'll make them as entertaining and fun as possible. Just roll with it, and it will make the time you spend with this book more enjoyable. Have fun, make jokes, and enjoy your new superpowers.

Learn Adobe Premiere Pro CC

When you're working on the projects in this book, you have the freedom to explore and make these projects your own. Of course, you're welcome to follow along with my examples, but please feel free to change text or styles to fit your own tastes. When you're sure you grasp the concepts I'm talking about, I encourage you to apply them in your own way. In some projects, you may want to even take things beyond the scope of what appears in the book. Please do so.

Prepare for the ACA exam

This book covers every objective for the Adobe Certified Associate (ACA) exam, but instead of organizing the book around the objectives, it's organized around workflows that you'll need to know on a real job, and the objectives are covered along the way. The authors of this series are teachers and trainers, and we've been doing this for a long time. We'll cover the concepts in the order that makes the most sense for *learning* and *retaining* the information best. You'll read everything you need to pass the exam and qualify for an entry-level job—but don't focus on that now. Instead, focus on having a blast learning Premiere Pro!

Develop your creative, communication, and cooperative skills

Aside from the actual hands-on work of learning Premiere Pro, this book explores the skills you need to become a more creative and cooperative person. These skills are critical for success—every employer, no matter what the industry, values creative people who can work and communicate well. This book describes the basics of creativity, how to design for (and work with) others, and project management.

Managing Files for Video Production ★ ACA Objective 2.4

Let's talk about some fundamental practices that are common to essentially all professional video production. These procedures help a production team stay organized and make it easier for any member of the team to manage and locate all of the media that's involved in a project.

▶ *Video 1.2* File Management Basics

Linking to files instead of embedding them

In other applications you might have assembled a document by pasting or importing text and graphics into it. This is also known as **embedding** imported content. When you save that document, its file size grows because it contains all the content you added. But embedding content is not practical for video projects, in part because video files themselves are very large. A single HD video clip can have a file size as large as thousands of text documents or photographs.

When you import media into a Premiere Pro project, the media is not copied into the Premiere Pro project file. Instead, Premiere Pro records the media's filename and folder path so it can retrieve the media from that location when it needs to display that media for you. The filename and folder path are the link to the media (**Figure 1.2**).

Figure 1.2 File path to linked media in Premiere Pro

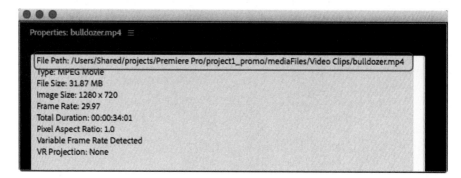

If you change the name of a file or move a file to a different folder, you break the link; Premiere Pro will lose track of the file and will be unable to load it. Fortunately, if that happens, Premiere Pro has tools that can help you quickly resolve links to lost files.

Because imported files are kept outside the project file, the project file size won't balloon as you add video. Another advantage is that if you need to swap in updated files for some that are already in the project, you have the option of simply replacing the old files with newer ones that have the same filename and location; Premiere Pro will simply pick up the newer ones.

But linking also means that you have the responsibility to make sure all the assets imported into a project are always accessible to the project. If you delete a linked video file that's used in a video project, the project will have a blank segment where that video used to be. Linking also means that you have the responsibility to make sure that when you create a backup of your project, you back up not just the project file but every file you imported. Naturally, that's easier to do if you've stored your files in an organized way.

Deciding where to store your files

When you use your computer, it's constantly responding to requests for file access from the operating system and from the applications you're using. For most applications, such as web browsers and word processors, the files that are accessed are relatively small and there are long breaks between reads and writes, so your computer has no problem keeping up with them.

But video production is different. As you edit video, and especially as you **scrub** through video looking for specific frames or checking your work, your computer continuously reads frame after frame after frame from your video files. It's basically reading (and sometimes writing) thousands of different pictures all the time, and this constant activity places unusually high demands on your computer. And there are many ways in which video editing can strain your system even more. For example, editing 4K video is much more demanding on your computer than editing 2K (1080p) video. The burden on your system is even higher if you're layering multiple video clips or applying image corrections or special effects to your footage. The more ambitious the project, the more difficult it is for your computer to keep up.

The unusually high performance requirements of video editing affect where you store video project files on your computer. If you store everything on the same drive, such as your main system drive, it's more likely that your computer will be unnecessarily slow while video editing. That's because the system and your video application will constantly be competing for access time on the same drive. Whenever either has to wait, you have to wait.

WORKING WITH MULTIPLE DRIVES

To avoid the performance problems associated with competing demands on one storage drive, video professionals spread out project files across multiple drives. Typically, the system drive stores the operating system and the video application (in this case Premiere Pro). But the media files that you're assembling into a project (video, audio, still images, etc.) are usually stored on a separate drive. Temporary working files that are generated during video editing, such as preview files and cache files, might be stored on a third drive.

The great advantage of distributing files across drives is that when the operating system needs access to its files and, at the same time, Premiere Pro requests access to video files and cache files, they aren't going to compete for the same drive. Now that each drive has just one job, it can more easily concentrate on maintaining its own data stream without interruption. You experience this as better responsiveness and smoother performance while editing video.

Splitting project files, media, and caches across drives is necessary when you use hard disk drives (HDDs). They have a set of heads that move together to retrieve files from the disk, and these mechanical heads are limited in how fast they can move from place to place on disk. When your data is on more than one drive, your computer can retrieve data faster, because now you have multiple sets of drive heads working simultaneously on different data transfers at the same time.

You may have heard that solid-state drive (SSD) storage is much faster than hard drive storage. That's true, and it's because SSDs don't rely on mechanical heads—they are solid-state memory modules. With no moving parts, SSDs can access large amounts of stored data at once. Although SSDs are more expensive than HDDs, they are so much faster that they can reduce the need to split project files across drives for performance reasons. But because newer formats such as 4K video are raising required data rates even more, distributing files across multiple SSDs is still a good way to help make sure your video editing system is as responsive as it can be.

What about network storage? Because of the high performance demands of video editing, it's not practical to store linked video on the most common types of network servers; the network transfer speed is too slow for real-time playback. There are network technologies that are fast enough, but they require such specialized and expensive equipment that you might encounter them only in a few high-end production studios.

COORDINATING ORGANIZATION WITH YOUR TEAM

How should you distribute the files across drives for your projects? If you work alone, you can decide for yourself based on your performance needs and your budget.

But in the projects you're working on for this book, you're working as part of a team. And that means you need to coordinate file organization with the production manager. When the company you're working for has established its own standard practices for organizing files, you need to follow them. These practices are typically set up so that a set of project drives can be passed among team members who all understand the agreed-upon organization of those drives. That way, if anyone on the team needs to work on the project, she can connect the drives to her computer and begin working without delay.

Logging and naming clips

When you edit a project that uses many video clips and other content files, to work efficiently you'll want to quickly find the files you need. Premiere Pro CC shows you thumbnail images of clips, but you'll often rely on filenames to pick out the correct clips to insert into the right parts of your production. You also don't want to waste time playing back bad clips in case they might contain footage you need. For these reasons, before you begin editing, you should perform a pass through all captured clips to delete bad takes and give each file a meaningful name.

You might put all of the initially available files in a pre-production folder and use that folder as a starting point for organizing the content into the project folders you'll actually use.

Managing project folders

To help make it easier to find the right content, it's often a good idea to keep different media types in different folders within your project folder. For example, you might want to keep all video clips in a Video Clips folder, all audio clips in an Audio Clips folder, and all still images and other graphics in a Graphics folder.

Consider adjusting the complexity of your folders to match the complexity of your project. For example, if you have many voiceover clips and background music clips, you may want to organize them into separate Voiceover and Music folders inside your Audio Clips folder.

It's also a good idea to have an Exports folder that serves as a destination for exporting your project drafts and finished video (**Figure 1.3**).

Figure 1.3 Project media organized with folders

Downloading, Unpacking, and Organizing

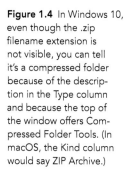

▶ **Video 1.3**
Unpacking and Organizing

For the projects in this book, you've been provided with a **ZIP file** containing the content you'll use. A ZIP archive provides a convenient way to combine multiple files into a single package that's easy to transfer online, so it's one way you're likely to receive project content.

The ZIP format is also popular for online transfers because it has built-in file compression; converting to ZIP can dramatically reduce the file size of some types of documents. However, many video and audio formats are already compressed, so adding those formats to a ZIP archive won't necessarily compress them any further.

DOWNLOADING LESSON FILES

For every project-based lesson, you'll need to download the lesson files from the Peachpit website. To do this, see "Accessing the free Web Edition and lesson files" on page xi.

UNPACKING A ZIP FILE

Although extracting content from a ZIP file works the same way in Windows and macOS, the results are slightly different.

In both platforms, simply double-click the ZIP file (**Figure 1.4**):

- Windows opens the ZIP file as a window that displays its contents. If you close the window, you still have the ZIP file.

Figure 1.4 In Windows 10, even though the .zip filename extension is not visible, you can tell it's a compressed folder because of the description in the Type column and because the top of the window offers Compressed Folder Tools. (In macOS, the Kind column would say ZIP Archive.)

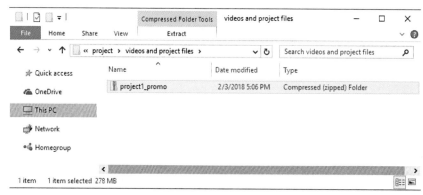

- macOS extracts the ZIP file into a new folder containing its contents. You now have both the original ZIP file and the new folder containing the contents of the ZIP file.

By default, the contents of a ZIP file unpack into the same folder that contains the ZIP file. In Windows, you can choose a different folder to save the unpacked files to by right-clicking the ZIP file and choosing Extract All. In macOS, you can unpack a ZIP file to a different folder if you use third-party utility software for opening and unpacking ZIP files.

> **NOTE**
>
> On a Mac, you may see several Thumbs.db files inside the lesson folders. You can delete all Thumbs.db files; they are included when a ZIP file is packaged in Windows but are not needed on a Mac.

ORGANIZING FILES INTO FOLDERS

With the ZIP file opened, you're ready to organize its contents into folders before you start editing. Video editing is typically team-oriented, so maintaining a consistent method for organizing and naming project folders makes it easier for team members to understand and use project files organized by another member of the team.

The project folder organization suggested in this book is just one way to do it; you might come up with your own way that works better for you. But don't get too attached to your own system because when you get hired by various production teams, you'll find that different teams have different standards for organization and file naming. Being adaptable in this area will make it easier for you to work with more teams.

In video 1.3, Joe shows how he consistently organizes project files into two folders called mediaFiles and pre-production. Within the mediaFiles folder, he creates the following subfolders to organize all of the source material that could go into a Premiere Pro project:

- **Audio Clips:** Use the Audio Clips folder to store audio-only files that will accompany the video, such as voiceover narration, music, ambient background sound, and sound effects.
- **Graphics:** Use the Graphics folder to store still images and artwork, such as photographs, logos, charts, maps, and icons.
- **Project:** Use the Project folder to store the Adobe Premiere Pro project file. You can also set Project Settings to store a project's cached files and rendered preview files into this folder.
- **Video Clips:** Use the Video Clips folder to store any motion clips, such as recorded video from video and still cameras, smartphones and tablets, aerial drones, and pre-rendered animations from other applications.

Now set up a hierarchy of properly named folders that you can use as a template for organizing each lesson:

1. In a folder window on the desktop, navigate to the folder that you've decided to use for storing all your Premiere Pro projects in the future.
2. Choose File > New Folder (or use whatever method you prefer for creating a new folder).
3. Name the new folder **Project Template**.
4. Open the Project Template folder, and create two new folders inside it named **mediafiles** and **pre-production**.
5. Open the mediafiles folder and create four new folders inside it named **Audio Clips**, **Graphics**, **Project**, and **Video Clips**.

You now have a complete folder organization template for a project. From now on, before you begin working on a new project, duplicate the Project Template folder and rename it for your new project, and after that one step you'll be ready to organize your project files.

Now let's organize the files for Project 1:

1. In the window containing the expanded contents of the Project 1 ZIP file, select the video files and then drag them to the Video Clips folder.
2. If the lesson files include audio-only files, repeat step 1 for any audio files, but move them into the Audio Clips folder.
3. If the lesson files include still graphics, repeat step 1 for any graphics files, but move them into the Graphics folder.
4. If the lesson files include any Premiere Pro project files, repeat step 1 for any graphics files, but move them into the Project folder. Some lessons don't include a project file because you create one during the lesson.

Finally, let's make an Exports folder for the finished videos that you render. Navigate to the folder that contains your Project Template folder, and create a new folder named **Exports**.

Joe stores the Exports folder at this level because it contains only the finished videos of each project he's worked on. It's one place where he can find all of his finished projects.

However, it's rare that the first export of a **sequence** is perfect; you'll often see something that you need to go back and correct and then export another version. If you don't want to clutter your top-level Exports folder with drafts, you might want

> **TIP**
>
> When organizing files, it's more efficient to drag more than one file at a time whenever possible. Use the multiple selection techniques of your operating system to do this; for example, with a folder displayed in List view, Shift-click a range of files to select multiple files for dragging.

to create an Exports folder inside each of your project folders as a place to collect draft versions as you evaluate them. When you export a version that you decide can be declared final, you can move that one up to the top-level Exports folder.

DELETING A ZIP FILE

After you unpack a ZIP file, you'll have two copies of the contents: the original downloaded ZIP file and the unpacked copies. You can now do one of the following:

- To free up some storage space, delete the ZIP file since you now have all the contents in unpacked form. You delete a ZIP file the same way you delete other files in your operating system; for example, drag its icon to the Recycle Bin (Windows) or Trash (macOS).
- If you're not concerned about storage space, you can keep the ZIP file around as a backup in case you want to start over with a project.

If the ZIP file is stored online, as it is with the lesson files for this book, you can safely delete the ZIP file after unpacking it since you can easily download it again if you need it. On the other hand, if your Internet connection is slow enough that downloading the project files takes a long time, you might keep the ZIP file around so that you don't have to download it again.

> **TIP**
> To see document types (video, audio, and so forth) more clearly, change a folder window to List view.

Identifying Job Requirements

During preproduction you should clearly describe the purpose, audience, deliverables, and other considerations for the promo video that need to be resolved before production begins. Let's take a look at the job requirements for Project 1:

- **Client:** Joe's Construction Cruisers. They use heavy equipment to prepare the land for large construction sites. Their tagline is "We Get the Dirty Jobs Done Right."
- **Target audience:** Joe's Construction Cruisers' clients are construction companies that work on large projects. They build housing developments, government buildings, shopping malls, schools, and large office buildings. They're typically males who are 30–60 years old.
- **Purpose:** To lead off an e-newsletter in a way that's interesting enough to maintain client attention and motivate more clients to open the newsletter and read its contents.

★ ACA Objective 1.1
★ ACA Objective 1.2
★ ACA Objective 1.3
★ ACA Objective 1.4

ACA Objective 2.4

*Video 1.4
Identifying Job Requirements*

- **Deliverable:** A 15-second video that features high-quality video clips and music with a positive, upbeat feel and clearly delivers the client's tagline. Technically, the video should be in H.264 YouTube 720p HD format, and it should be compressed so that it loads quickly online while still retaining acceptable image quality.

Listing available media files

In this project, some media has already been acquired for the project. What do you have to work with?

- Aerial clips from a construction site
- Rights-cleared music
- Company logo from the client
- Voiceover audio files

That set of media is sufficient to complete the job, so you don't need to acquire any more media. Editing can begin.

Preventing potential legal issues

You can shoot anything you see, but you can't necessarily use it. Reproducing a likeness of a person or personal property and using it to promote a business can interfere with the legal and privacy rights of people and property owners. To avoid unwanted legal consequences, be sure to obtain all necessary permissions and clearances through model and property releases.

For this promotional video, licensing concerns are addressed as follows:

- The aerial clips are covered by location releases from the property owners and model releases for any recognizable people.
- The music from the stock library is **licensed** for distribution in this type of production, and the music has been approved by the client.
- The client's logo is covered by an artist release.
- The voiceover is covered by a model release for the recognizable voice.

Be mindful of other media too. Watch out for things like music unintentionally recorded as part of the ambient audio of a location video clip, or a copyrighted poster on the wall of a room in a scene. Carefully review "free clip art" or "free photo" websites too, because some websites contain work that has not been properly released or is being redistributed without the permission of the creator.

If you're not sure whether a media item is legal to use, consult a lawyer familiar with video production. Lawyers know what to look for, and they can recognize media that may cause legal problems for you later. Another reason it's a good idea to work with a lawyer is that requirements can vary; for example, legal requirements for commercial productions may be stricter than for editorial uses. Copyright law is specific to your country.

Starting Premiere Pro

★ ACA Objective 2.2

▶ **Video 1.5** *Start Premiere Pro*

You start Premiere Pro just as you start any other application you use, but what might be a little different is what Premiere Pro presents to you immediately after it starts up.

When you start Premiere Pro for the first time, the Welcome screen appears—a travelogue-like full-screen video that introduces the application (**Figure 1.5**). Three buttons appear over the video: Get Started, Watch, and Explore. These aren't covered in this course, so you can look through them on your own time if you want. To continue with this lesson, click Skip.

Figure 1.5 The Welcome screen in Premiere Pro

TIP

If you want to reset Premiere Pro preference settings when you start Premiere Pro, press and hold hold the Alt (Windows) or Option (macOS) key immediately after Premiere Pro begins opening.

The Welcome screen should not appear again unless you reinstall Premiere Pro or reset its preferences, but if you want to see the Welcome screen again in the future, choose Help > Welcome.

> **TIP**
>
> You can also start Premiere Pro from the Adobe Creative Cloud desktop application or by typing its name into Windows desktop search or macOS Spotlight search.

To start Premiere Pro:

1. Do one of the following:
   - In Windows, click the Adobe Premiere Pro CC application icon on the Start menu, Start screen, or Taskbar. If a shortcut icon for Premiere Pro exists on the desktop or in a folder window, you can double-click that.
   - In macOS, click the Adobe Premiere Pro CC application icon in the Launchpad or Dock. If an alias icon for Premiere Pro exists on the desktop or in a folder window, you can double-click that.
2. Choose an option from the Start screen (**Figure 1.6**).

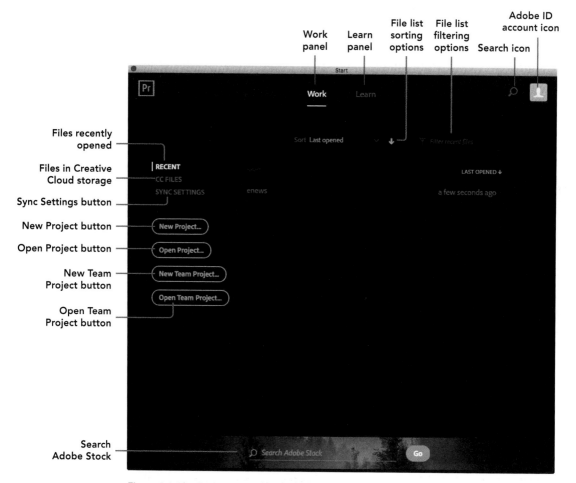

Figure 1.6 The Start screen in Premiere Pro

Unlike many applications, Premiere Pro doesn't show you a blank workspace when it launches. Instead, you see the Start screen, which is designed to help you begin working in or learning about Premiere Pro. As a beginner, you might take advantage of the Learn tab for tutorials. As an intermediate user, you might use the New Features and Tips & Techniques tabs to get caught up with the latest enhancements.

If you're a beginner, consider clicking the Learn tab to watch tutorials that teach you how to make different kinds of Premiere Pro documents.

When you're ready to make a new Premiere Pro document, the Start screen gives you several ways to get started in the Work tab. The main way is to click the New Project button, which is a shortcut for the File > New > Project command. That opens the New Project dialog box, where you enter the specifications for a project; we'll look more closely at that soon.

If you'd like to start from a template, type what you're looking for into the Adobe Stock search field at the bottom of the Start screen and click Go, which will take you to an Adobe Stock web page that may contain graphics, video, and Premiere Pro templates relevant to your search term. (Adobe Stock may not be available on computers in some schools and organizations.)

In day-to-day use, the first thing you'll often want to do after starting Premiere Pro is open an existing project so you can continue working on it. When the Start screen appears, make sure the Work tab is active and click Recent, which displays a list of Premiere Pro projects that were opened recently. (The Recent list may be empty if this is your first time using Premiere Pro on this computer or if Premiere Pro preferences were reset.) The Recent list is a shortcut for the File > Open Recent command.

If the project you want to work on isn't in the Recent list, click the Open Project button. That's a shortcut for the File > Open Project command, and it works like the Open command in other applications.

> As in other applications, if you prefer to use keyboard shortcuts, you can create a new file by pressing the keyboard shortcut for the File > New > Project command, which is Alt+Ctrl+N (Windows) or Option+Command+N (macOS). Or you can open a file by pressing the keyboard shortcut for File > Open Project, which is Ctrl+O (Windows) or Command+O (macOS).

If you have Premiere Pro documents in Creative Cloud Files online storage, you can see them by clicking CC Files. Creative Cloud Files is cloud storage associated with the Adobe ID that's signed into the computer. Creative Cloud Files works much

> **NOTE**
>
> If you're using a shared computer where you can't sign in with your own Adobe ID, Creative Cloud Files storage might not be available to you.

the same way as other cloud storage services you may have used, such as Dropbox, Google Drive, Microsoft OneDrive, or iCloud Files. You can transfer files to and from Creative Cloud Files storage using a web browser, folders on your Windows or Mac desktop, or a mobile app on your phone or tablet.

The Start screen includes other options that are outside the scope of this book because they're more advanced. The Sync Settings button is for using your Creative Cloud account to synchronize your Premiere Pro settings, such as preferences, with other installations of Premiere Pro. New Team Project and Open Team Project are buttons you'll use only if you're part of a workgroup using the cloud-based Adobe Team Projects collaborative video workflow for Adobe Premiere Pro CC, Adobe After Effects CC, and Prelude CC. If your organization is interested in using Team Projects, see

www.adobe.com/creativecloud/team-projects.html

As useful as the Start screen is, when starting Premiere Pro you might prefer to have the application automatically open the most recently used project instead. To change this preference, open Premiere Pro Preferences, and in the General pane, choose the option you want from the At Startup drop-down list.

Setting Up the New Project Dialog Box

★ ACA Objective 2.1

If you're the type of person who clicks OK as soon as a dialog box appears, you don't want to do that with the New Project dialog box. That's because it contains three tabs of settings that define fundamental aspects of your video project, including where the project and its working files are stored. Although it's possible to change New Project settings after you've started working on a project, it's much easier and better to set New Project settings mindfully and correctly the first time.

Configuring the General tab

The most important thing you'll do in the General tab of the New Project dialog box (**Figure 1.7**) is set the name and storage location for the project. A project can contain several named video sequences, so you generally want to name the project after the overall work (such as the title of a feature-length movie) and later name the sequences in the project after sub-sections of the work (such as scenes within that movie).

Figure 1.7 The General tab of the New Project dialog box

With each new project, double-check Location as well as the name. If you don't change the Location, Premiere Pro will save the project in the folder where the previously created project was saved, which may not always be what you want.

1. For Name, type **Promo**. You don't need to enter a filename extension; it will be added automatically.

2. Click Browse, and then set project Location to the folder you created for the promo video, where you unpacked and organized the promo project files. In video project 1.5, Joe chooses a location on a second storage drive that he uses as a data drive, separate from the drive containing his operating system and applications.

3. For the Renderer option under Video Rendering and Playback (**Figure 1.8**), choose a **Mercury Playback Engine** GPU Acceleration option when possible. This option attempts to accelerate Premiere Pro performance on your computer, using hardware such as the graphics processing unit (GPU).

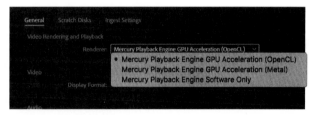

Figure 1.8 When the Renderer option gives you a choice, choose a GPU Acceleration option.

Chapter 1 Introduction to Adobe Premiere Pro CC   19

If a **CUDA** option is available, that is typically the fastest option. However, it's best to test all the GPU Acceleration options to see which one is the fastest, because the answer is not always the same. If two computers have different combinations of graphics hardware, CPU speed and number of cores, and amount of RAM, the fastest GPU acceleration method may be different for each of those computers.

Mercury Playback Engine Software Only is typically the slowest option, so use it only if a problem prevents you from completing your project with a GPU Acceleration option on or if it's the only option because GPU Acceleration isn't supported on the computer you're using. If you know your system has a supported GPU and Premiere Pro is still not recognizing it, check to see that you have the latest GPU drivers installed.

ACCELERATING PERFORMANCE WITH THE MERCURY PLAYBACK ENGINE

The Mercury Playback Engine is a set of technologies that Adobe developed to make video editing faster and more responsive whenever possible. Whether you have the Mercury Playback Engine set to Software Only or GPU Acceleration, it accelerates your work by coordinating and making the best use of 64-bit CPU processing, multithreaded CPU processing, RAM, and fast storage for performance caching. Acceleration is more effective with larger amounts of installed RAM, more CPUs, more free space available on scratch drives, and faster drives (such as SSDs instead of HDDs).

Selecting the GPU Acceleration option usually results in much faster rendering. It can enhance performance even more using powerful graphics card technologies such as **OpenCL**, **Apple Metal**, and CUDA. The performance benefits of GPU Acceleration are so dramatic that video professionals specifically choose graphics cards that support it.

If the GPU Acceleration option is not available, it means your computer has a graphics card that doesn't meet the system requirements for the Mercury Playback Engine. You may need a graphics card that is newer or more powerful. Adobe maintains a list of compatible graphics cards on its website:

https://helpx.adobe.com/premiere-pro/system-requirements.html

4   When editing video from digital cameras, leave the Video and Audio Display Format options at their default settings: Timecode and Audio Samples, respectively.

5. Capture Format matters only if you will be capturing clips from video equipment in this project; it doesn't apply if you're copying video files from media cards or other drives. Set Capture Format to HDV for video projects involving capture of high-definition digital formats. The DV option is for older standard-definition digital video formats.
6. You may want to select "Display the project item name and label color for all instances" as Joe does in video 1.5.
7. Configure the other tabs as needed (see the following sections), and click OK. The new project opens.

Configuring the Scratch Disks tab

How you configure the **Scratch Disks** tab depends on how many drives you can attach to your computer. As you learned in "Working with Multiple Drives" earlier in this chapter, video editing is faster when the demands of accessing large amounts of video data are spread out across multiple drives, because multiple drives can transfer that data in parallel. But because there are some situations where you end up working with a single drive, we'll cover that too.

If you're working with a team, consult them before setting up the Scratch Disks tab because the team may have specific requirements for scratch disk locations. For example, some may require that the project be stored on an external drive with scratch files set to Same As Project so that when the project drive is handed to another team member, all the scratch files stay with the project instead of being left behind on different computers.

SETTING UP A ONE-DRIVE SYSTEM

If you're using a computer with only one storage drive available, such as a laptop, you can leave all the Scratch Disks options set to Documents or Same As Project. Performance may not be optimal, but since there are no other drives to use, it's the only option.

SETTING UP TWO DRIVES

If you're using a second drive to store your project and media files, as Joe does in the video, you can set up your scratch drives this way:

1. In the New Project dialog box, set Location to the second drive.

2. Click the Scratch Disks tab (**Figure 1.9**).

Figure 1.9 The Scratch Disks tab of the New Project dialog box

3. Choose Same As Project for all of the drop-down lists.

SETTING UP THREE OR MORE DRIVES

If you're able to connect more than two drives to your computer, you might organize your Scratch Disks connections in the following way:

- **Location:** Set Location to the drive where you want to keep project files, such as in the data drive example in video 1.5.

- **Captured Video** and **Captured Audio:** If you'll be capturing media from video equipment with this project, click Browse and choose a location on a drive that is used only for video capture so that it does not have to handle competing read/write tasks. This will help prevent dropped frames when capturing. If you have enough drives, you can set Captured Video and Capture Audio to different drives to further lower the risk of dropped frames.

- **Video Previews** and **Audio Previews:** Click Browse and choose a location on a fast drive dedicated only to storing rendered previews.
- **Project Auto Save:** It's okay to set Project Auto Save to Same As Project because project files are small, quick and easy to save, and saved only every few minutes.
- **CC Libraries Downloads** and **Motion Graphics Template Media:** You can use these types of components across several projects, and they don't take up much space or impact performance very much, so you may want to choose Documents, a central location on your system drive. If they need to travel with the project, you might choose Same As Project. If they need to be stored in a special location for a workgroup, click Browse and choose that location.

WHAT KIND OF DRIVES AND CONNECTIONS SHOULD YOU USE FOR SCRATCH DISKS?

If more than one internal drive can be installed inside your computer, install additional drives in unused drive bays (preferably SSDs), and then assign those drives as scratch disks in Project Settings in Premiere Pro.

If your computer doesn't have any unused internal drive bays, external scratch disks are most effective when they're connected using your computer's fastest port. USB 3 ports are common and reasonably fast, whereas USB 3.1 ports are faster and preferred. Though less common, the fastest external storage ports available on high-end computers today use the Thunderbolt standard; Thunderbolt 3 is the fastest version.

Whether internal or external, connecting SSD storage is preferred if you can afford it, because SSDs are much faster than hard drives. If you must connect external hard drives, USB 3 ports will be more than fast enough for them.

Avoid using USB 2 or USB 1 to connect scratch drives, because they are much slower than current standards.

> **NOTE**
>
> *If you're using a computer with only one drive, don't partition the drive to simulate a multiple-drive system. That won't create any performance advantage because it's still a single-drive mechanism trying to fulfill multiple data requests at once, and free drive space will be more limited. You get the performance benefits of setting up multiple scratch disks only when the computer is connected to multiple actual drives.*

ABOUT CC LIBRARIES

For some projects, if you have access to Creative Cloud Libraries, you may find it useful to set the Creative Cloud (CC) Libraries location in the Scratch Disks tab. If you're using Creative Cloud on a school or library computer, it might not have access to Creative Cloud Libraries or other services, but if you have your own individual Creative Cloud membership, you probably do have it. CC Libraries are cloud-synced sets of content that can be shared among members of a team or among your mobile devices and your computer. For example, if another member of your team is using Adobe Photoshop to develop graphics for your video production, that person can add them to a named CC Library that you can open from Premiere Pro so that you can import those graphics into your video project. Or, you can use the Adobe Capture CC mobile app to sample real-world colors and add them to a CC Library. You can then load that CC Library into Premiere Pro so you can add those colors as a color-grading look in your project.

About the Ingest Settings tab

The **Ingest** Settings tab is useful when you want to preprocess every video clip that you import into this project. For example, some production workflows require that all clips be *transcoded*, or converted, to a specific format for editing. Ingest Settings options can also create *proxies*—small placeholder versions of original clips—for faster editing when the computer you're using may have trouble keeping up with the processing demands of high-definition video. Proxy workflows are useful when the video resolution is very high (such as 4K resolution and up) or when the computer is older or less powerful.

You won't use the Ingest Settings tab in this lesson.

Locating a Project and Editing Its Settings

There may be times when you're not sure of the folder or even the drive where an open project is stored. Fortunately, the folder path to a Premiere Pro project file is always listed in the title bar for the application window (**Figure 1.10**).

Figure 1.10 The Project path in the application window title bar; the asterisk at the end means there are unsaved changes.

If you need to change the settings you entered in the New Project dialog box, choose File > Project Settings and then choose either the General or Scratch Disks command.

If you're trying to change settings such as Frame Rate and Frame Size, those are settings for a sequence, not a project. A project can contain multiple sequences with different settings. To change the settings for individual sequences within a project, choose Sequence > Sequence Settings when a sequence is selected or active.

GETTING HELP AND LEARNING NEW SKILLS

To learn more about a task you're trying to accomplish, click the Help menu and choose Adobe Premiere Pro Help or Adobe Premiere Pro Support Center.

The Help menu isn't the only place to get assistance with Premiere Pro issues. You can also get help from various online sources:

- To ask other users questions on the Adobe Communities forum for Premiere Pro, go to https://forums.adobe.com/community/premiere.
- To get news about Premiere Pro, check the Premiere Pro CC blog at https://theblog.adobe.com/creative-cloud/premiere-pro/.
- To get updates about Premiere Pro on Twitter, follow @AdobePremiere.
- To get Creative Cloud support on Twitter, follow @AdobeCare.

Exploring Panels and Workspaces

★ ACA Objective 2.2

 *Video 1.6 Exploring Panels and Workspaces*

The Premiere Pro CC user interface is designed to be powerful and flexible enough for professional workflows. Like other applications you may have used, it has menu commands with keyboard shortcuts as well as floating panels of options that you can arrange.

In Premiere Pro, you'll spend most of your time using controls in panels. If you can't find the control you're looking for, the panel that contains it might be hidden. To open any panel, including the hidden ones, click the Window menu and choose the name of the panel you want (**Figure 1.11**).

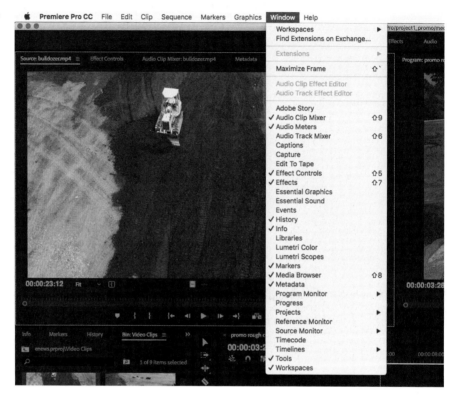

Figure 1.11 The Window menu in Premiere Pro CC lists all available panels, including ones that might be hidden in your current workspace.

Looking at the primary panels for video editing

If you haven't used a professional video editing application before, the panel arrangement in Premiere Pro may be unfamiliar, but it has a long history—it's based on the traditional window layout for video editing. You'll also find this panel layout in other video editing applications because it reflects how dedicated video monitors were arranged in an edit bay in the days of tape-based video editing.

In the Editing workspace, the three largest panels in the application window are the Source and Program panels across the top half and the **Timeline** panel across most of the bottom half (**Figure 1.12**). They are open when a sequence is open in a project. The Timeline and Program panels are two views of a sequence; the Timeline panel shows how the video clips and other content are arranged in time within the sequence, and the Program panel shows the current state of the video at a specific time in the sequence. As you play back a sequence, the playhead (a vertical line in the timeline, also called the Current Time Indicator) moves along the Timeline, and the Program panel displays the video at the time indicated by the playhead.

Figure 1.12 The Source, Program, and Timeline panels in their default arrangement

At the bottom-left corner is the Project panel. This contains all the media that's been imported into the project, including media that you haven't yet used in a sequence.

Video editing typically involves the following traditional video editing workflow. If you want your test project to have content for you to try out for the rest of this chapter, you can follow these steps:

1 View a video clip in the Source panel, usually by double-clicking it in the Project panel. Trim that clip's beginning and end in the Source panel if needed.

2   Add the clip to the timeline, either by dragging and dropping or by using a keyboard shortcut. Clips have to be added to a sequence to appear in the timeline. If a sequence isn't already open in the timeline, dragging a clip to the timeline creates a sequence based on that clip.

3   See the results in the Program panel, and replay the clip if needed to review whether the resulting program plays back as expected.

This cycle of moving from the Project panel to the Source panel and then to the Timeline and Program panels is fundamental to assembling a video program. You'll use it to build your program from beginning to end by adding clip after clip to the timeline.

Exploring important panels

A video sequence isn't just about video clips in a timeline. It will often include audio, effects applied to the video, and other content such as still images and **titles**. The Premiere Pro user interface includes panels that let you work with all of those:

- The Project panel (**Figure 1.13**) contains all media imported into the project, as you've seen before. Sometimes a project has a very long list of media, which can make it challenging to locate a specific file, so you can create named **Bins** inside the Project panel. Bins in the Project panel work much like folders on your desktop; you can nest bins inside bins just as you can nest folders inside folders.

- The Effects panel (**Figure 1.14**) includes lists of effects and **transitions** for video and audio along with color grading presets.

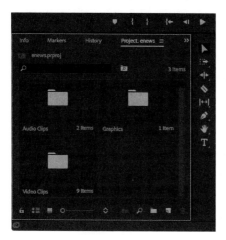

Figure 1.13 Project panel

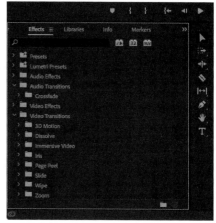

Figure 1.14 The Effects panel, which may be grouped with the Project panel

- The Tools panel (**Figure 1.15**) contains tools that you can use to edit and view media interactively. Notice that many of the tools have a tiny triangle in the bottom-right corner; this indicates that additional tools are hidden behind that tool. To see the hidden tools, click and hold the mouse button on the tool, and the tool group pops out.
- The Audio Meters panel (**Figure 1.15**) shows you the audio level at the playhead. It includes clipping indicators so that you can instantly see whether audio levels are too high.

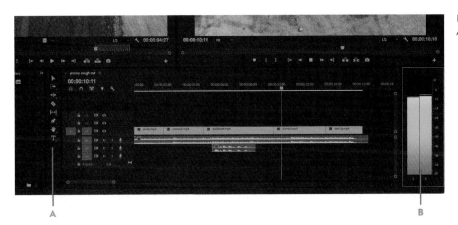

Figure 1.15 Tools (A) and Audio Meters (B) panels

Arranging panels

The first time you start Premiere Pro, you'll see all panels contained within a single application window. Panels share dividers with adjacent panels, so making one panel smaller makes another panel larger.

Panels can be arranged in three ways (**Figure 1.16**):

- **Docking**: You can attach panels to each other along their edges. This lets you view sets of panels side by side so you can see more information at once.
- **Grouping**: You can combine panels so that they share a single space, like a stack of paper folders. When panels are grouped, you can see their named tabs along the top of the group like folders in a file cabinet. If you've used tabbed windows in a web browser, tabs in panel groups work the same way.
- **Floating**: You can pull a panel out of a set of docked or grouped panels so that it's positioned on its own, in front of other panels.

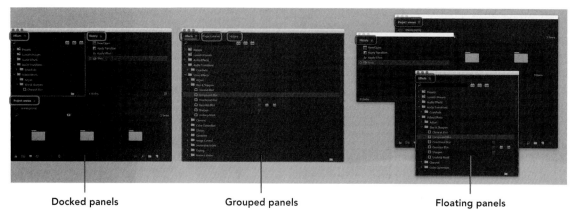

Figure 1.16 You can arrange panels in several ways.

When many panels are grouped in a narrow space, there might not be room to show all of the panel tabs. You can click an overflow menu (**Figure 1.17**) to see the names of all of the panels in that group. If you want to close a panel, there are two ways to do that. If a panel is grouped with others, click the panel menu and choose Close Panel; if the panel is floating, click the Close button in the upper-left corner (**Figure 1.17**).

Figure 1.17 Window management controls

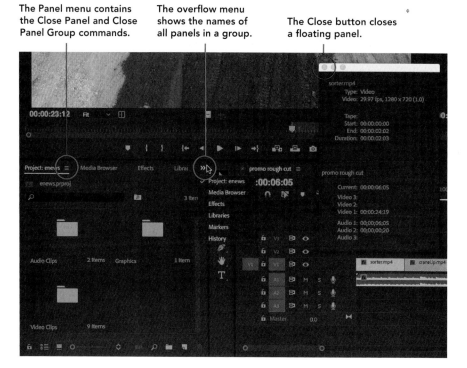

30 Learn Adobe Premiere Pro CC for Video Communication

DOCKING A PANEL

Let's try changing a grouped panel into a docked panel:

1 Click the panel menu in the Effects panel tab, and choose Close Panel.

2 Choose Window > Effects.

You're able to reopen the Effects panel because all panels are listed under the Window menu.

The Effects panel is currently grouped with other panels, so it can end up behind other panels in its group. Suppose you'd like to see the Effects panel all the time, because you plan to use a lot of effects. You can drag it out of the group and dock it instead.

3 Drag the Effects panel tab, and then, without releasing the mouse button, hold the panel over the other panels in its group. Notice the drop zone overlays that temporarily appear over a panel when you drag another panel over it and how the drop zone under the pointer is always the darkest. One of three things will happen (**Figure 1.18**):

- If you drop a panel over a center drop zone, Premiere Pro will group the panel you drop with the panel you drop it onto.
- If you drop a panel over an edge drop zone, Premiere Pro will insert the panel you drop next to that edge of the panel you drop it onto.
- If you press and hold the Ctrl (Windows) or Command (macOS) key while releasing the mouse button, the panel will become a floating panel.

Figure 1.18 The drop zone you use determines how a panel attaches to others.

4 Position the pointer over the right drop zone of the panel group, and when it's highlighted, release the mouse button.

The Effects panel is now docked between its former panel group and the Tools panel (**Figure 1.19**).

Figure 1.19 Dropping the Effects panel on the right, edge drop zone of a panel group inserts the Effects panel to the right of the panel group.

5 Position the pointer over the right edge of the Effects panel, which is also a vertical divider between the Effects panel and the Tools panel.

6 Drag horizontally to adjust the width of the two panels (**Figure 1.20**). As you make one panel larger, the adjacent panel becomes smaller. Adjust the divider for the best use of space; for example, you might want to make the Effect panel wider and the Tools panel narrower. You can drag vertical and horizontal dividers.

Figure 1.20 Dragging a vertical panel divider

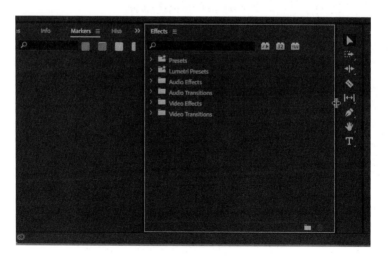

FLOATING A PANEL

Sometimes a panel is easier to work with if it's completely independent, floating above other panels. Let's float the Effects panel:

1   Click the panel menu in the Effects panel, and choose Undock Panel (**Figure 1.21a**).

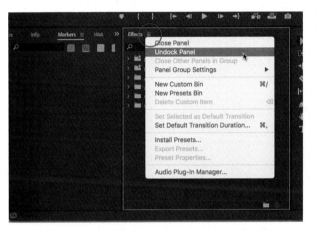

Figure 1.21a Undocking the Effects panel

> **NOTE**
>
> *Undocking is what the tutorial video demonstrated as floating a panel, though in the video it's done with a shortcut. In Windows, Ctrl-drag a panel tab; in macOS, Command-drag a panel tab.*

2   Drag the title bar of the Effects panel to position it anywhere on the screen (**Figure 1.21b**).

Figure 1.21b Drag the title bar to move the undocked panel.

> **TIP**
>
> *The application window doesn't have to fill the screen. You can resize it by dragging any corner or side. Making the application window smaller can be useful if you want to reveal the desktop or another application behind Premiere Pro so that you can drag and drop items between applications.*

3   If you want to resize the panel, position the pointer over any panel edge or corner until you see a two-headed arrow, and drag.

Other panels can be docked and grouped with this floating panel.

GROUPING A PANEL

It's easy to put a floating or docked panel into a panel group.

Drag the Effects panel tab, position it over the center drop zone of the panel group containing the Projects panel, and release the mouse button to drop it (**Figure 1.22**).

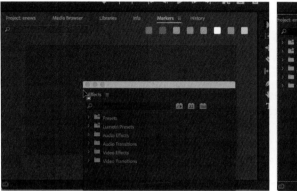 

Figure 1.22 Dropping the Effects panel into a group

This is the same panel group that the Effects panel was originally part of.

As in other applications that use tabbed windows, you can arrange the order of tabs in a panel group by dragging them left or right within the group.

> ### ARRANGING PANELS ACROSS MULTIPLE MONITORS
>
> If you want to use Premiere Pro across multiple monitors, you can move some panels to the second monitor. Simply drag the tab of any panel to the second monitor. As soon as you drag the panel outside the application window, it will become a floating window just like the application window. Position this second window on the second monitor.
>
> You can now drag any other panels into the Premiere Pro window on the second monitor, docking and grouping them into an arrangement just like you can in the main application window.

Using Workspaces

An arrangement of panels is called a *workspace*. You can choose from one of the preset workspaces that present an optimized selection of panels for specific tasks such as sequence assembly or color correction. You can also arrange panels the way you want and save the layout you've created as your own named workspace. You can see the available workspaces and switch the current workspace using the Workspaces panel across the top of the application window and on the Window > Workspaces menu.

★ ACA Objective 2.2

When you open Premiere Pro for the first time after installation, it uses the Editing workspace (**Figure 1.23**). You can see this because the Editing workspace is highlighted in the Workspaces panel at the top of the screen.

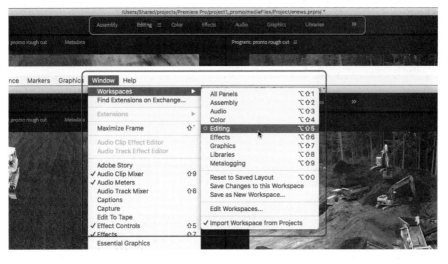

Figure 1.23 You can choose and edit workspaces using the Workspaces panel and the Window > Workspaces menu.

In the Workspaces panel, click a few of the other workspaces to see how they change the panel layout, and then click the Editing workspace.

Premiere Pro remembers all the changes you make to the panel arrangement across sessions, so after you exit, the next time you start the application it will restore the panel arrangement you were using in the previous session. But sometimes you want to revert to the original panel arrangement before the end of an editing session. The panel arrangement is called a *workspace*, and you can reset the workspace to restore its original arrangement.

To discard changes you've made to the current workspace, choose Window > Workspaces > Reset To Saved Layout. If the Workspaces panel is visible, you can also click the menu icon to the right of a workspace's name and choose Reset To Saved Layout (**Figure 1.24**).

Figure 1.24 Choosing Reset To Saved Layout

Now let's customize the panel layout in a way that's useful for upcoming lessons and save it as a workspace:

1 Close the Info, Libraries, and Markers panels. Because they're grouped, close them by clicking each of their panel menus and choosing Close Panel.

The panel changes you're making in this exercise are just suggestions. In your own work, you can decide which panels you want open and how you want to arrange them.

2 Choose Window > Workspaces > Save As New Workspace.

3 Enter a name for the new workspace (**Figure 1.25**); for this lesson name it **Simple Editing**, and click OK.

Figure 1.25 Saving a new workspace

Notice that the new Simple Editing workspace you created has been added to both the Workspaces panel and the Workspaces submenu. One of the reasons workspaces are listed in two places is because the Workspaces panel may be hidden at times or in some custom workspaces.

Using Premiere Pro on a PC or Mac

▶ **Video 1.7** Using Premiere Pro on a PC or Mac

Premiere Pro is designed to work essentially the same way whether you're using it on a PC running Microsoft Windows or an Apple Mac computer running macOS. There are only a few minor differences between using Premiere Pro on a PC or Mac, so even if your computing experience is mostly based on one or the other, it should be relatively easy for you to do Premiere Pro jobs for clients running either operating system.

Before you dive into video editing, it's good to become familiar with how Premiere Pro relates to the conventions of the platform you're using.

Using keyboard shortcuts

PCs and Macs use different sets of **modifier keys**—the keys you press to change how features or other keys work:

PC	Mac
Ctrl	Command
Alt	Option
Shift	Shift

The Mac has a Control key in addition to the Command key, but it is not frequently used except to emulate a right-click (see "Using context menus" later in this chapter).

> ### BACKSPACE/DELETE AND FORWARD DELETE: SIMILAR BUT DIFFERENT KEYS
>
> A PC (Windows) keyboard has a large Backspace key at the top-right corner of the keyboard, and an extended keyboard has a separate, smaller forward delete key (usually marked Del). In some applications, including Premiere Pro, these keys do different things. A Mac keyboard has a large Delete key at the top-right corner of the keyboard, and an extended keyboard has a smaller, separate forward delete key (confusingly marked Delete); those can also be programmed to do different things. When you study keyboard shortcuts, be sure you understand whether you should be pressing the Backspace/Delete key or the forward delete key.
>
> Compact keyboards, such as on laptop computers, might have only the Backspace (PC) or Delete (Mac) key at the top-right corner of the keyboard. If you need the function of the forward delete key, you can usually get it by also pressing the Fn key. For example, on a compact Mac keyboard you can use the Delete key as a forward delete key by pressing Fn-Delete.
>
> It's common to accidentally confuse the Backspace/Delete and forward delete keys, so if you're trying to use a shortcut that involves the Backspace/Delete key and it isn't working, see if the shortcut is supposed to use the forward delete key instead.
>
> If you're not sure whether a Premiere Pro shortcut uses the Backspace/Delete or **forward delete** key, check the list of its keyboard shortcuts posted online by Adobe:
>
> https://helpx.adobe.com/premiere-pro/using/default-keyboard-shortcuts-cc.html

Using context menus

A PC comes with a two-button mouse or trackpad, so the standard way to open a **context menu** is to click with the right mouse button.

A Mac comes with a mouse or trackpad that may only perform a single click by default, but you can configure it to recognize a right-click (called a secondary click) in the Mouse or Trackpad panel in System Preferences. You can also connect a two-button mouse to a Mac. Another way to open context menus is to press and hold the Control key while clicking the primary mouse button.

On a PC or Mac, you can also connect other input devices such as a trackball or graphics tablet with a stylus, and you can typically set up one of their extra buttons to act as a right-click.

Opening Preferences

As you become more familiar with Premiere Pro, you may find yourself going into the Preferences dialog box to make Premiere Pro fit your working style, your hardware configuration, and the needs of your specific production workflow. The Preferences command appears under different menus on a PC and Mac, and Premiere Pro follows the convention of each platform.

To open the Preferences dialog box on a PC, choose Edit > Preferences.

To open the Preferences dialog box on a Mac, choose Premiere Pro CC > Preferences.

★ ACA Objective 2.4

Video 1.8
Importing Media

Importing Media

Premiere Pro is similar to other professional media editing and management tools in that you don't copy and paste content into it; you import files into it. Importing makes it possible for Premiere Pro to record the folder path to the content's source file so that the file can be linked.

You'll import media into the Project panel (**Figure 1.26**). The Project panel is the central repository for all media that you've either used in your project or set aside for potential use in your project.

Bins are nested subsections of the Project panel that you can break off and float independently of the panel. In this way, bins are a lot like subfolders on your desktop. You can work with media in bins the same way that you can in the Project panel.

Figure 1.26 Project panel containing imported media and bins

Importing in Premiere Pro is flexible enough to let you bring in media using the way that is most productive for you. You can drag and drop files directly into a project, or you can use menu commands or keyboard shortcuts.

1 Open the enews project.
2 Make sure the Project panel is visible.
3 Import media using any of the following methods:
   - Drag and drop individual files into the Project panel. This may be easier if you resize the Premiere Pro application window so that you can see the desktop window that you want to drag from.
   - Drag and drop folders into the Project panel (**Figure 1.27**). Each folder becomes a bin in the Project panel.

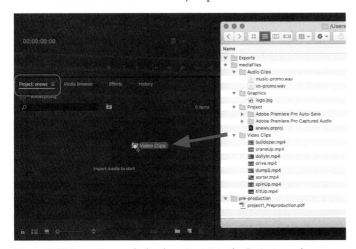 

Figure 1.27 Importing media by dragging it into the Project panel

Chapter 1 Introduction to Adobe Premiere Pro CC   39

- Choose File > Import, select an image, and click Import.
- Choose File > Import, select a folder, and click Import. The folder becomes a bin in the Project panel.
- Press the keyboard shortcut for the Import command: Ctrl+I (Windows) or Command+I (macOS).
- Double-click inside Project panel or bin. This opens the Import dialog box.

Use any combination of these techniques to import the video, audio, and graphics files for Project 1.

Organizing imported media

Immediately after importing, make sure the files you imported are productively organized in the Project window. For the projects in this book, it's good to have Project window bins that correspond to the different folders on the desktop that were created to organize the various types of media used for the project, such as Video Clips, Audio Clips, and so on.

To create a new bin, click the New Bin button at the bottom of the Project window, or choose New > Bin (**Figure 1.28**).

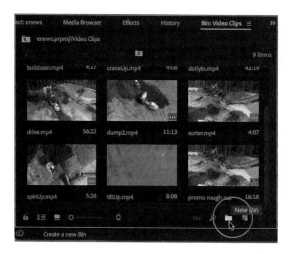

Figure 1.28 Create a bin by clicking the New Bin button.

To organize files with bins, drag and drop anything in the Project panel into bins, just as you would when organizing files into folders on your computer desktop.

As Joe demonstrates in video 1.8, you can change how a bin opens. Open the Preferences dialog box, and in the General pane, adjust the Bins settings to match the behavior you want.

Viewing imported media

The Project panel isn't just for storage. You can use it to preview, arrange, and trim media.

★ ACA Objective 4.1

Similar to your computer desktop, you can see imported media as a list or as a grid of icons.

The Thumbnail view of the Project panel is useful for previewing and arranging your content:

- Preview the contents of a video clip by hover-scrubbing: Move the pointer horizontally over a video thumbnail image (**Figure 1.29**).
- As you hover-scrub a clip, you can trim it. To set an **In point**, press the I key; to set an **Out point**, press the O key. The blue line under the thumbnail indicates the time range between the clip's In point and Out point.
- Drag to arrange clips in the order in which you'd like them to appear when you create a sequence.

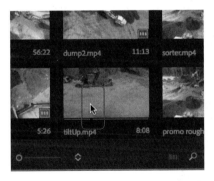 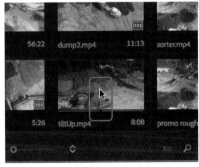

Figure 1.29 Hover-scrubbing is a quick way to preview a video clip.

List view is useful when you want to inspect information details about your media (**Figure 1.30**). You can do the following:

- See bins as a hierarchy and expand them to see their contents.
- Sort the list by clicking a column heading.

TIP

Be sure to scroll List view all the way to the right to see all of the columns that are available, such as the column that lets you mark media as Good.

TIP

Want to import sequentially numbered still images as a single video clip, such as a time lapse? Choose File > Import, select the first image, and select the Image Sequence option at the bottom of the Import dialog box.

- Change the column order by dragging column headings horizontally.
- Customize which metadata columns are displayed by choosing Metadata Display from the Project panel menu.

Figure 1.30 List view can provide you with detailed information about your clips.

IMPORTING ALTERNATIVE: THE MEDIA BROWSER PANEL

Don't like going through the Import dialog box again and again? Without leaving Premiere Pro, you can use the **Media Browser** panel to browse your computer and all of the volumes connected to it, including network volumes and media cards. When you locate media using the Media Browser panel, you can import it in one step by simply dragging it into the Project bin, or you can select it in the Media Browser panel and choose File > Import From Media Browser.

Whether or not you use the Media Browser, before you import media always make sure you first copy it to a drive that will be connected. This highlights a critical difference between the Media Browser and the Project panel: content that you see in the Media Browser is not necessarily imported into any project, whereas content you see in the Project panel is definitely imported into that project.

Understanding a Basic Editing Workflow

ACA Objective 4.1

In video 1.9, Joe uses a simple salad analogy to explain a basic video editing workflow (**Figure 1.31**):

▶ **Video 1.9**
Understanding a Basic Editing Workflow

- The Project panel is like a refrigerator where you keep the ingredients for your salad. In the same way that you can organize items inside a refrigerator using bins, you can use bins to organize media items in the Project window.
- The Source panel is like a kitchen cutting board. In the same way that you would trim off end pieces of vegetables so that only the good parts go into in the salad, you can open video and audio clips into the Source panel and trim off the ends of each clip so that only the good part in the middle goes into the video sequence.

Figure 1.31 If you can make a salad, you can edit a video.

1. Store video ingredients in the Project panel (refrigerator).
2. Trim ingredients in the Source panel (cutting board).
3. Add and mix ingredients in the Timeline panel (salad bowl).
4. Realize the finished dish in the Program panel (plate).
5. Export the plate to the dinner table.

- The timeline is like a salad bowl. In the same way that you would move vegetables from a cutting board into a salad bowl, you can move trimmed media clips from the Source panel into the Timeline panel. In the Timeline panel, you mix video clips together to form a sequence. In the same way that you could add dressing and croutons to make the salad more interesting, in the timeline you can also add non-video enhancements such as titles, music, and graphics.

- The Program panel is like the plate. Whereas the timeline is an abstract graphic display of the media items sequenced in time, the Program panel shows you the visual content itself. In the Program panel you can play back the current state of the sequence to see how close it is to the finished video you want to create.

- Exporting is like serving the finished salad to the dinner table. Although you can preview the sequence in the Program panel, it's still just a sequence of separate media files. Exporting fuses all of the elements together into a single video file, and at the same time exporting can compress the file to a size appropriate for your final delivery medium.

In the basic video editing workflow, you start from the Project panel, trim clips in the Source panel, sequence them in the Timeline panel, preview them in the Program panel, and export the sequence. As you gain experience, you'll learn other ways to perform these steps that may be faster or more productive, but for now, the salad analogy is a good representation of how many video workflows work.

Editing a Sequence

★ ACA Objective 3.1

▶ *Video 1.10*
Working with Sequences

What's a sequence? It's a timeline within a project. The basic use of a sequence is to organize clips in the correct order.

A sequence isn't the same as a project because a project can contain multiple sequences. For example, you might edit a movie by creating one sequence for each scene and then take the edited sequences and add them all to one master sequence where you put together all the scenes.

Being able to use multiple sequences also means you can create sequences with different specifications in a single project. For example, a project can contain multiple versions of the same video program as different sequences. If you're editing a long movie, one sequence might be the full-length, full-resolution version edited and color-graded for a digital cinema projector in a movie theater, whereas another sequence might be a version of the same content edited and graded for viewing on a home television. You might also have sequences that are a two-minute theatrical trailer and a 30-second television commercial derived from the main sequence.

In a typical sequence workflow, you take clips from the Source Monitor or a bin, add them to a sequence to create a **rough cut**, and then use established editing techniques to fine-tune the edits until the sequence has the pacing and flow you want. If the program has a specific duration requirement, such as a 30-second TV commercial, part of the goal of editing is telling the story properly while fitting all the required clips within that time limit.

Let's get the enews project ready to work with sequences:

1 If you've added any sequences to the enews project, delete them from the Program panel now so that the project contains only the media files you imported earlier. Sequences are marked by a sequence icon ( ).

2 If you're viewing a bin, switch to the top level of the Project window. You can go up one level by clicking the button to the left of the folder path (**Figure 1.32**); keep clicking until you reach the top level (enews.prproj).

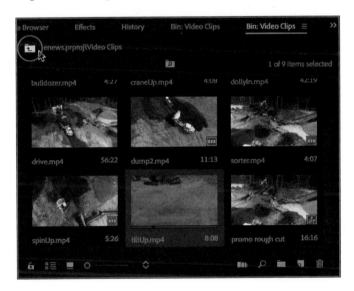

Figure 1.32 To move up a level in the Project panel, click the button next to the folder path.

Creating a new sequence

When creating a new sequence, you must ensure that the specifications of the sequence—such as frame rate, aspect ratio, and **pixel aspect ratio**—are appropriate for mastering. That doesn't necessarily mean that the specs should match the final export specifications. For example, a project may need to be output in multiple formats for targets such as HDTV, smartphone, and the web.

Premiere Pro is designed so that it's perfectly okay to mix media types in a single sequence, so don't be afraid to mix clips from a 4K cinema camera, a 1080p camcorder, a smartphone, and an action camera. You might want to set the sequence itself to the highest-quality format you expect to deliver.

> **TIP**
> The clip you use to create a sequence will become the first clip in the new sequence, so it's best to use the clip that you actually want to appear first.

But how do you do that? If you create a new sequence from nothing, the Sequence Settings dialog box can be intimidating because it contains many options labeled with technical terminology. Fortunately, you may not need to set those options individually because Premiere Pro provides a simple shortcut: you can create a sequence based on a video clip, and the sequence will automatically be based on the technical specifications of that clip. For example, if you know you want to create a sequence that fully supports the 1080p footage from your camcorder, you can create a sequence based on one of its clips. There's more than one way to do that.

1 In the Project panel or a bin, do one of the following:

- Select a clip and choose File > New > Sequence From Clip.
- Drag a clip to the New Sequence button in the Project panel or bin.
- If the Timeline panel is empty, drag and drop a clip from the Project panel or bin into the Timeline panel (**Figure 1.33**).
- Right-click (Windows) or Control-click (macOS) a clip and choose New Sequence From Clip.

Figure 1.33 One way to create a new sequence based on the specifications of that clip is to drag a clip to an empty Timeline panel.

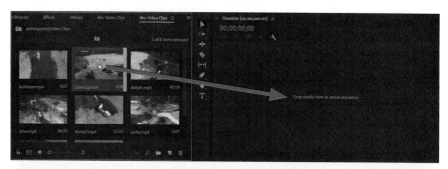
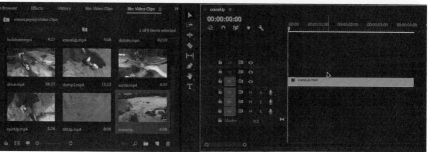

46   Learn Adobe Premiere Pro CC for Video Communication

2. In the Project panel, look for the new sequence, identified by a sequence icon ( ) to distinguish it from the clip of the same name that it was created from.

3. Click to highlight the name of the new sequence, and rename it **promo rough cut** (Figure 1.34).

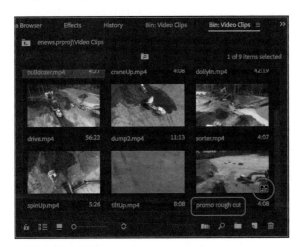

Figure 1.34 The new sequence renamed, displaying its sequence icon

Renaming a new sequence is important because a sequence created from a clip takes on the clip name. For example, if you want the final name of the video to be **promo rough cut**, it's a good idea to give the sequence that name, because it will become the default filename for the exported file later.

The new sequence was created in the same bin as the clip it was based on, so it's in the Video Clips bin. Let's organize the project a little further by taking the sequence out of the Video Clips bin:

1. Make sure the Project panel is at the top level (not in any bin), and switch it to List view.

   Using List view lets you see items at multiple levels of the project window simultaneously, although you could also do this by floating the Video Clips bin away from the Project panel or putting it in a different panel group.

2. Click the New Bin button to create a bin for storing sequences, and name it **sequences**.

Chapter 1 Introduction to Adobe Premiere Pro CC 47

3   Expand the Video Clips bin, and drag the promo rough cut sequence into the sequences bin (**Figure 1.35**).

Figure 1.35 Create a bin for sequences.

NOTE

If the first clip you add to a sequence does not match the sequence settings, Premiere Pro will ask you if you want to change the sequence settings to match the clip; clicking Change Sequence Settings is often a safe choice. But if your sequence settings are based on specifications that must be maintained, click Keep Existing Settings.

You may want to verify that the sequence settings are correct. There's more than one way to do this for the selected sequence:

- As Joe demonstrates in video 1.10, you can choose Preview Area from the Project panel menu so that you can compare the specifications displayed at the top of the Project panel when you select the sequence and then select the clip it was based on. The Preview panel shows only basic sequence settings.
- Choose Sequence > Sequence Settings. The resulting dialog box shows all sequence settings.

TIP

There are times when it's useful to maximize one of the panels. To do this without permanently modifying the workspace, position the pointer over a panel and press the tilde (~) key. Press the tilde key again to restore the panel to its usual place in the workspace.

TIP

If you've already arranged clip thumbnails in the order you want in the Project panel or a bin, you can create a sequence from those clips in one step. Select the clips and choose Clip > Automate To Sequence, or click the Automate To Sequence button in the Project panel or bin.

Building a rough cut

★ ACA Objective 4.1

▶ **Video 1.11** Create a Rough Cut

A rough cut is like a first draft of a video program. You can use it to make sure that the basic concept behind the program works well, such as the order of clips and the total running time. For a rough cut, you don't have to obsess over the finer details of timing and editing.

USING THE INSERT AND OVERWRITE BUTTONS

Earlier, you learned about In and Out points and how they can be set for clips in the Project panel or a bin. You can set In and Out points more precisely using the Source Monitor.

First, let's start from scratch so that we can add trimmed clips to the timeline. If craneUp.mp4 is in the promo rough cut sequence, select it and press the Delete key.

1. In the Project panel, double-click a clip. It opens in the Source Monitor.
2. Use the Source Monitor playback controller to find the frame where that clip's action should start.
3. Click the Mark In icon (**Figure 1.36**), or press the I key.

Figure 1.36 You'll find the Mark In icon at the bottom of the Source Monitor.

4. Use the Source Monitor controls to find the frame where that clip's action should stop.

5 Click the Mark Out icon, or press the O key (**Figure 1.37**). The time ruler now displays a gray segment that represents the duration between the In and Out points you set. This is the duration that will be added to the sequence.

Figure 1.37 The Mark Out icon is just to the right of the Mark In icon.

6 In the Timeline panel, make sure the playhead is at the time where you want to add the clip that's open in the Source Monitor. In this case, make sure it's at the beginning of the timeline.

7 Do one of the following (**Figure 1.38**):

- To add the clip at the playhead so that its duration moves all following clips later in time, click the Insert button or press the comma (,) key.
- To add the clip at the playhead so that it replaces any clips that occupy its duration, click the **Overwrite** button or press the period (.) key.

TIP

Before adding clips with keyboard shortcuts, make sure the sequence you want to work on is active in the timeline.

In this example, there are no other clips on the timeline, so it doesn't matter which option you use.

The clip is added to the timeline. Also, the playhead automatically moves to the end of the clip you added, ready for you to insert or overwrite another clip there next.

NOTE

If the clip is added to a different track than the one you were expecting, check the blue rectangle in the first column in the Timeline panel. The track containing that rectangle is the one where clips are added by Insert or Overwrite.

Figure 1.38 Add a trimmed clip from the Source Monitor to the Timeline panel by clicking the Insert button in the Source panel.

8   Go back to step 2 to add another clip at the playhead time, and repeat steps 3–8 until you've added around four or five clips to the sequence (**Figure 1.39**).

Figure 1.39 Adding more trimmed clips to the sequence

Chapter 1   Introduction to Adobe Premiere Pro CC   **51**

Figure 1.40 Use the video track-targeting buttons in the Timeline panel to change which track receives media added from the Source panel.

When you use the Insert and Overwrite buttons, clips are added to the timeline on the track that's currently targeted. The track targeted for receiving source media is the one marked by a blue rectangle in the first column of the Timeline panel, at the left edge. By default, V1 is the target track (**Figure 1.40**), but you can change that by clicking a different track in the first column in the Timeline panel.

If you add video that has audio, you'll see that both video and audio tracks are added to the Timeline panel, and they're linked to each other. If you want to edit a clip's video and audio separately, select the clip and choose Clip > Unlink.

CREATING A SEQUENCE BY DRAGGING AND DROPPING MEDIA

Another way to build a sequence is to drag and drop the clips instead of using the Insert and Overwrite buttons. To assemble your sequence that way, after setting In and Out points for a clip in the Source panel, drag the clip from the Source panel to the timeline and drop it on the track you want at the point in time where you would like to add it.

When you build a sequence by dragging, a normal drop is like using the Overwrite button. If you'd rather drag as if you were using the Insert button, press and hold the Ctrl key (Windows) or the Command key (macOS) as you drop the clip into the timeline.

> **NOTE**
> In time notation, the numbers after the last colon are frames. For video, you read time as hours, minutes, seconds, and frames.

As you build your sequence, you'll want to become familiar with the control layouts in the Source Monitor and Program Monitor (**Figure 1.41**).

Navigating time quickly is important for editing efficiently. Anywhere you see a Playhead Position time display in Premiere Pro, you can navigate time using these techniques:

- Drag the playhead.
- Use the Program panel transport control buttons, or their keyboard shortcuts, to move forward or backward along the timeline.
- Scrub the Playhead Position time display by dragging the pointer horizontally over it.
- Click the Playhead Position time display, type the time to which you'd like the playhead to move, and press Enter or Return.

> **TIP**
> You can enter time quickly by entering only the numbers without the leading zeros. For example, to move the playhead to 00:01:29:03, enter 12903.

A Playhead Position time display
B Magnification
C Buttons for adding markers and In/Out points
D Drag audio only or video only to Timeline
E Playhead transport controls
F Insert, Overwrite, Export Frame buttons
G Playback resolution
H Settings icon

Figure 1.41 Important controls in the Source Monitor and Program Monitor panels

You can use keyboard shortcuts to go to the edit points between clips. To go to the previous edit point, press the Up Arrow key. To go to the next edit point, press the Down Arrow key.

You'll start over in the next section, so when you're finished experimenting in this section, delete all the clips from the Timeline panel:

1 Make sure the Timeline panel is active (outlined with a highlight); if it isn't, click it.

2 Choose Edit > Select All.

3 Do one of the following:

- Press the Delete key.
- Choose Edit > Clear.
- Right-click any selected clip and choose Clear.

★ ACA Objective 4.1

 **Video 1.12** Create a Rough Cut from the Project Panel

Creating a rough cut from the Project panel or bins

It's possible to assemble a rough cut directly from the Project panel or bins, bypassing the Source panel, because you can do some basic trimming in the Project panel or in bins.

Earlier you learned that you can set In and Out points in the Project panel. This is one of the techniques Joe demonstrates in video 1.12. Try this now:

1 In the Project panel, navigate to the Video Clips bin and make sure it's set to Icon view.

2 Hover (don't drag) the pointer over the video thumbnail of a clip you haven't added yet, moving it left and right to scrub the clip's frames.

3 Hover-scrub to find the frame where you want to set an In point, and press the I key.

4 Hover-scrub to find the frame where you want to set an Out point, and press the O key.

The blue line under the thumbnail icon indicates the range between the In point and the Out point, which will be added to the sequence.

5 Drag the trimmed clip to a track in the Timeline panel (**Figure 1.42**).

Figure 1.42 Drag the trimmed clip to the timeline.

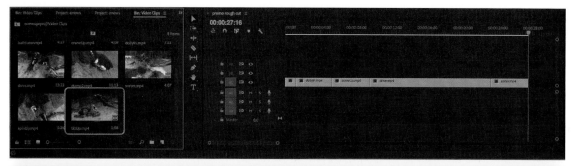

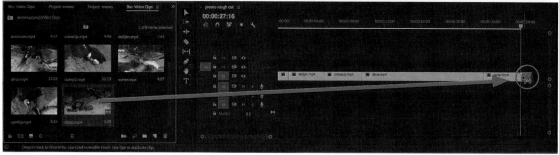

54 Learn Adobe Premiere Pro CC for Video Communication

Because the main reason to go through the Source panel is to trim clips before adding them, dragging clips straight from the Project panel to the timeline makes the most sense when the clip you're dragging is already perfectly trimmed and does not need further work in the Source panel. This is more common with audio files. You'll add those to the timeline from the Project panel now:

1. In the Project panel, navigate to the Audio Clips bin.
2. Drag music-promo.wav from the Audio Clips bin to the beginning of the A1 track in the timeline.
3. Drag vo-promo.wav from the Audio Clips bin about 5 or 6 seconds into the A1 track in the timeline (**Figure 1.43**).

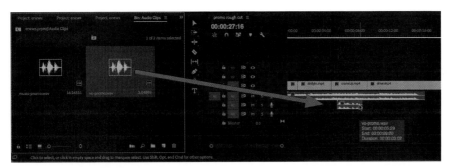

Figure 1.43 Dragging an audio clip into the A1 track

4. Choose File > Save to save the project.

BUILDING A ROUGH CUT WITH KEYBOARD SHORTCUTS

Many experienced video editors prefer to work much faster by using only the keyboard. Premiere Pro is designed to accommodate this style of working so that you can use it when you feel that you're sufficiently familiar with the editing workflow. For example, you can perform the following functions using keyboard shortcuts:

- Switch to the Project panel or bin by pressing Shift+1.
- Select a clip in the Project panel or bin with the arrow keys.
- Open the selected clip in the Source Monitor by pressing Shift+O.
- Navigate to the clip in the Source Monitor by pressing L or the spacebar to play, K to stop playback, and J to play in reverse. Those keys are next to each other on the keyboard, so the "JKL" set of shortcuts was designed to form a reverse/stop/play set of shortcut keys that you can use to navigate video with three fingers.

continues on next page

continued from previous page

> **TIP**
>
> When you use the JKL keyboard shortcuts for shuttling through a clip or sequence, pressing J or L more than once increases playback speed. For example, pressing LL or LLL results in higher speeds.

- When you find your In point and Out point frames, press I or O, respectively.
- Press the comma (,) key to insert the clip into the program at the playhead in the timeline, or press the period (.) key to overwrite. Those shortcuts are designed as an adjacent pair right under the JKL keys.

If you like to work this way, you can study a complete list of keyboard shortcuts here:

https://helpx.adobe.com/premiere-pro/using/default-keyboard-shortcuts-cc.html

The keyboard-driven editing workflow in Premiere Pro is based on shortcuts that have become standard over many years in professional video editing suites. That means you can generally use similar or the same shortcuts, such as the JKL set, to edit quickly on other professional video editing systems you might use.

Navigating the Timeline

★ ACA Objective 2.3a

▶ **Video 1.13** Navigating the Timeline

In Premiere Pro you'll probably spend the bulk of your time using the Timeline panel, so it's good to be familiar with the layout of its controls. You can experiment with these techniques using the enews project you've been working on.

> **TIP**
>
> The \ (backslash) key is very useful in the Timeline panel, because it toggles between the current magnification and fitting the entire sequence within the visible width of the Timeline panel.

★ ACA Objective 4.1

▶ **Video 2.8** Work in the Timeline

The main area of the Timeline panel is where you see audio and video tracks and the clips and other media sequenced. Here are some of the most useful ways you'll navigate the Timeline panel (**Figure 1.44**):

- Zoom the time magnification level using the scroll bar at the bottom, the Zoom tool (🔍), the + (plus)/– (minus) keyboard shortcuts, or the backslash key (\) so that you can see the entire duration between the first and last clips in the sequence.
- When the timeline is magnified, change which time range is visible by dragging the middle of the scroll bar at the bottom or by using the Hand tool (✋).

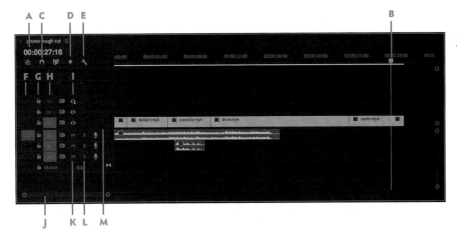

Figure 1.44 Overview of Timeline panel controls

A Playhead Position time display
B Playhead
C Snap
D Add Marker (to sequence, not a clip)
E Timeline panel settings
F Track source targeting
G Lock track
H Track labels (Video 1, Video 2, Audio 1, Audio 2...)
I Show/hide video track in Program panel
J Time magnification
K Mute audio track
L Solo audio track
M Video/audio track separator

- Move the playhead in time by dragging it, entering the time into the Playhead Position time display, or using keyboard shortcuts such as the spacebar to play/pause your video playback.
- Control which tracks are targeted, locked, hidden, muted, and so forth by using the icons along the left side of the Timeline panel.

TIP

If your keyboard has Home and End keys, you can use them to navigate a sequence. To go to the first frame of a sequence, press Home. To move to the last frame, press End.

Exploring the Editing Tools

Let's take a quick look at some fundamental tools and techniques for editing video in the timeline. You can experiment with these techniques using the enews project you've been working on, but make sure that in the timeline, the V1 track has four or five video clips on it. If it doesn't, add them now. To configure your timeline to work best in this section, do the following:

- Make sure the clips you add to the timeline are trimmed (have an In point and Out point set), because several editing tools depend on having extra material before the In point and after the Out point of a clip.
- Add enough clips to span the length of the longer audio track (music-promo.wav).

★ ACA Objective 4.1
★ ACA Objective 4.3

▶ Video 1.14
Reviewing the Editing Tools

Figure 1.45 The Tools panel

You'll want to get to know the Tools panel (**Figure 1.45**), because most of the tools are there to help you handle different editing situations.

Remember, the tiny triangle at the bottom-right corner of a tool means other tools are hidden behind that tool (**Figure 1.46**). You can see and select the hidden tools by holding down the mouse button on the tool in the Tools panel.

This section is just an introduction that you can use to start editing quickly. Many of the tools introduced here are covered in more detail in later chapters and videos.

TIP

If the mouse isn't doing what you expect, check which tool is selected. Many common edits are done with the Selection tool, so you may get unexpected results if a more specialized tool is selected.

Figure 1.46 Tool groups in the Tools panel

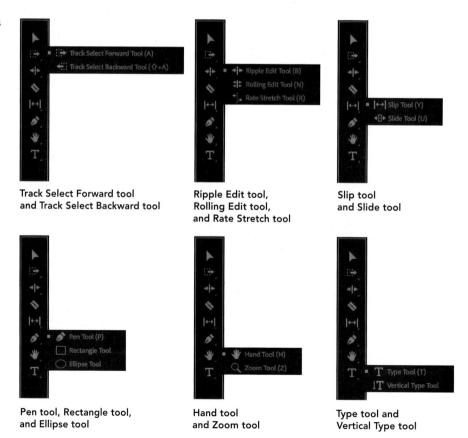

Using the Selection tool

You'll probably spend most of your time in Premiere Pro with the Selection tool (![icon]) active. This is partly because it's the tool you want to use for arranging clips on the timeline. But the Selection tool is also useful because it often adapts to the task at hand. For example, if you position the Selection tool over one end of a clip in the timeline, it automatically changes to a Trim pointer so that you can drag to trim a clip without changing tools. Let's try this:

1. Select the Selection tool, and in the timeline for the enews project, click a clip. The clip highlights to let you know it's selected (**Figure 1.47**).

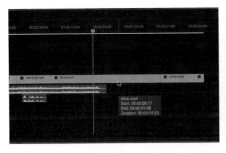 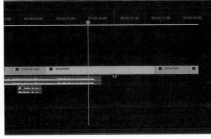

Figure 1.47 Clicking a clip selects it.

2. With the Selection tool, drag a selection rectangle around several clips on the timeline. Premiere Pro highlights those clips to let you know they're selected (**Figure 1.48**).

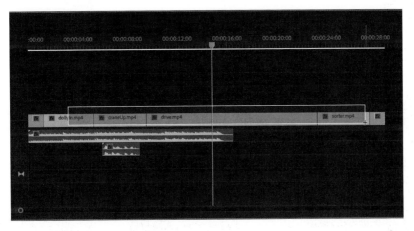

Figure 1.48 Dragging a selection rectangle around several clips selects them.

If you were to apply a command, option, or effect that affects clips, the selected clips would be altered. You can also drag selected clips with the Selection tool to move them to a different part of the timeline.

TRIMMING WITH THE SELECTION TOOL

Trimming clips in the Timeline panel is different from trimming in the Source Monitor. In the Source Monitor, you trim the clip by itself. In the Timeline panel, you trim while being aware of how a clip interacts with the clips that are before and after it in the sequence and how an edit affects the total duration of the entire sequence. These factors affect which technique you choose to make a specific edit.

1. To get ready to trim a clip, in the Timeline panel move the playhead to the part of the clip you want to trim, and zoom into that time so that you can see the trim point frames precisely and clearly.

2. Choose an editing technique:

    - To trim the end of one clip without altering the rest of the sequence in any way, use the Selection tool to drag the end of a clip (**Figure 1.49**). This may leave a gap between clips.

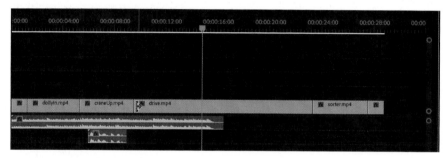

Figure 1.49 Trim a clip by using the Selection tool to drag the left or right end of the clip.

    - To trim the end of one clip without leaving a gap by automatically shifting all following clips in time, hold down the Ctrl key (Windows) or Command key (macOS) while using the Selection tool to drag the end of a clip (**Figure 1.50**). This makes the Selection tool work as the **Ripple Edit** tool, which you'll learn more about later in this section. A ripple edit shifts the times of clips after the edit point so that the edit does not leave a gap.

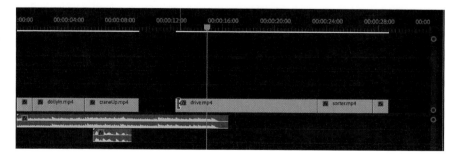

Figure 1.50 Ctrl/Command-dragging the Selection tool trims a clip using a ripple edit.

Notice that as you drag the end of a clip, the Program panel displays a preview of the resulting frames on both sides of the edit point you're adjusting (**Figure 1.51**). You'll see this type of interactive preview with other editing tools too.

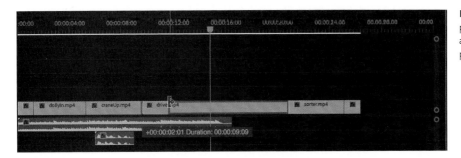

Figure 1.51 The Timeline panel previews the frames around a trim as you perform it.

In the Timeline panel you fine-tune the trimming of each clip until the sequence has the right order, pacing, flow, and duration. Another way to look at trimming is that it's an adjustment to the In or Out point of clips in the timeline.

Keep in mind that trimming is nondestructive. In and Out points are merely markers that affect when playback of a clip starts and stops; they don't delete frames from the original clip. If you make a clip shorter by setting a later In point, you can subsequently restore the lost frames by setting an earlier In point for that clip.

Using the Ripple Edit tool

A ripple edit is useful whenever you want to trim the beginning or ending of a clip without leaving a gap between it and an adjacent clip. You don't need to use a ripple edit to trim the last clip in a sequence because there aren't any clips after it, so you can simply drag the end of the last clip with the Selection tool. But when you want to trim at any other edit within a sequence, including the beginning, you'll probably want to use the Ripple Edit tool. Because a ripple edit affects all clips after the edit point, a ripple edit changes the duration of the sequence.

To use the Ripple Edit tool (![]), select it in the Tools panel, and then in the timeline, use the Ripple Edit tool to drag the end of a clip (**Figure 1.52**).

Figure 1.52 Before and after a ripple edit

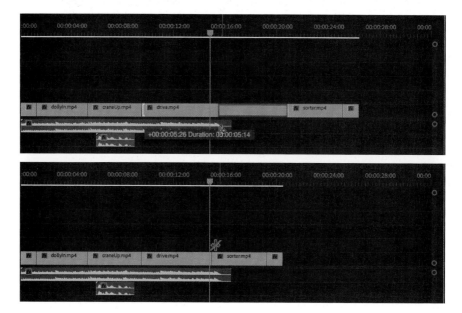

Using the Rolling Edit tool

A **rolling edit** is useful whenever you want to trim the edit point between two clips without leaving a gap and without changing the total sequence duration. If a rolling edit doesn't change those things, what does it change? A rolling edit changes the Out point of the clip before the edit point, and the In point of the clip after the edit point. On the timeline, what this looks like is that the time of the edit point is shifted, affecting only the two clips on either side of the edit point.

To use the Rolling edit tool (![]), select it in the Tools panel, and then in the timeline, use the Rolling Edit tool to drag the end of a clip (**Figure 1.53**).

When trimming with a tool, pay attention to the appearance and direction of the pointer as it approaches the edge of a clip. Make sure the areas that highlight indicate that the tool will trim the clip you intend.

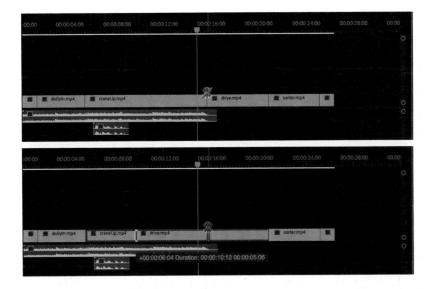

Figure 1.53 Before and after a rolling edit

Using the Razor tool

When you want a clip to start or end at a specific frame, you can simply drag one end with the Selection tool. But you can also use the Razor tool ( ) to split the clip at that frame.

To use the Razor tool, select it in the Tools panel, and then in the timeline, click a clip at the frame where you want to split it (**Figure 1.54**). The clip becomes two instances that you can edit separately, or you can delete one.

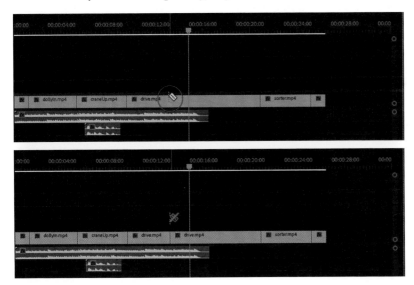

Figure 1.54 Before and after using the Razor tool

> **TIP**
>
> To instantly extract a still frame from a clip or sequence, click the Export Frame button in the Program panel controller.

The new edit point created by the Razor tool becomes the Out point of the first instance and the In point of the second instance, so splitting a clip with the Razor tool should not look any different on playback until you edit the resulting instances.

Using the Slip and Slide tools

The difference between the **Slip** tool ( ) and **Slide** tool ( ) is similar to the difference between the Ripple Edit tool and the Rolling Edit tool. Whereas the Ripple Edit and Rolling Edit tool are different ways to change the timing of an edit point, the Slip tool and Slide tool are different ways to change the timing of an entire clip.

Use the Slip tool when the only things you want to change are the In point and Out point of one clip in the sequence. When you move a clip forward or backward in time with the Slip tool, the clip that changes is the one you're dragging; its In point and Out point both shift by the number of frames you drag. The rest of the sequence doesn't change, and the duration of the clip you dragged doesn't change.

Use the Slide tool when you want to move a clip forward or backward in time without changing its In point and Out point. That clip's duration doesn't change, and the total duration of the sequence doesn't change. What does change are the Out point of the clip preceding the clip you dragged, and the In point of the clip following the clip you dragged.

A simple way of thinking of Slip and Slide is that Slip is like shifting a clip behind adjacent clips, whereas Slide is like shifting a clip in front of adjacent clips; that's what it looks like when you use each tool in the timeline.

To use the Slip tool or Slide tool (**Figure 1.55**), select it in the Tools panel, and then in the timeline, use the tool to drag one clip forward or backward in time along its track.

Figure 1.55 Comparing the way the Slip tool and Slide tool affect clips in the timeline

Rearranging clips in a sequence

In most cases, you can rearrange clips using the Selection tool on its own or with a modifier key. In some cases you may need to use a more specialized tool. Depending on what you want to do, rearrange timeline clips using one of the following methods (**Figure 1.56**):

- To move one or more clips to a different time, select them with the Selection tool and drag them.
- To move one or more clips to a different time while automatically shifting clips that the moved clips would otherwise overwrite, select clips with the Selection tool and Ctrl-drag (Windows) or Command-drag (macOS) them. This performs a ripple move.
- To quickly select multiple clips on all tracks after a certain time, click the Track Select Forward tool () in the timeline at the time where you want selection to start. The Track Select Backward tool does the same thing with tracks before the time at which you click the tool.

Pay attention, because corresponding audio tracks will also be selected.

> **TIP**
> If a tool affects tracks you don't want it to affect, click to enable the Lock icon (🔒) for the tracks you don't want to alter. The Lock icons are in the second column from the left side of the timeline.

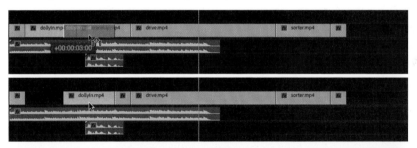

To move a clip in time, drag a clip with the Selection tool.

To move clips in time using a ripple edit, Ctrl/Command-drag a clip with the Selection tool.

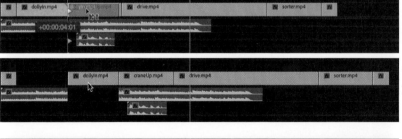

To move all clips on a track after a certain time, use the Track Select Forward tool.

Figure 1.56 Three ways to change the timing of clips in the Timeline panel

Working with Audio

★ ACA Objective 3.1

★ ACA Objective 3.2

★ ACA Objective 4.7

▶ Video 1.15
Working with Audio

Audio editing is an essential part of most video productions. Fortunately, you'll find that basic audio editing is somewhat similar to basic video editing: you can arrange audio clips in the Timeline panel, trim their ends, fade them in and out, and apply effects to clips or tracks.

As you work with audio, always be aware of whether you are working with an entire audio track or just an audio clip. For example, it's possible to apply an effect to a clip or a track; if you apply it to a clip, the rest of the track won't have that effect.

If you're following along with the videos using the enews project that Joe is working with in the videos and you have also locked or muted some audio tracks while editing video earlier, before continuing with this lesson make sure you have unlocked and unmuted all audio tracks.

Looking at Timeline panel audio controls

In the Timeline panel, you'll find audio controls along each track; you have seen some of them in use in the videos already. To the left of each audio track, these are the controls that differ from the video track controls (**Figure 1.57**):

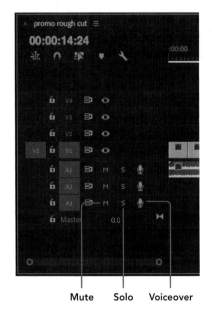

Figure 1.57 Audio controls in the Timeline panel

- **Mute:** Enable Mute on a track to silence that track.
- **Solo:** Enable Solo on a track to play only that track, silencing other tracks.
- **Voiceover:** Click the Voiceover button on a track to start recording on that track, and then begin speaking. Click it again to stop recording; a new audio clip with your speech appears on the track. If the voiceover isn't recorded properly, right-click the Voiceover icon, choose Voice-Over Record Settings, and verify that the settings are appropriate for your system (such as the selected audio source).

Other audio controls can be revealed on the audio clips that are on the timeline. You'll work with those later in this section.

Making audio waveforms and controls easier to see

Working with audio often requires adjusting audio controls that appear over the audio clips on the timeline. These controls, and the visual representation of the audio waveform, are easier to see if the audio track is taller. You can make the audio tracks taller in these ways (**Figure 1.58**):

- Click the wrench icon (🔧) near the top-left corner of the Timeline panel and choose Expand All Tracks.
- In the timeline, in the area containing controls just to the left of the track area (not in the track area), position the pointer over the bottom edge of the audio track you want to adjust, and drag downward to increase the height of that track.
- When editing many audio tracks, you may want to allocate more vertical space to the audio tracks. In the area containing controls just to the left of the track area (not in the track area), position the pointer over the double line dividing the video and audio sections of the timeline, and drag up to give more space to the audio tracks.

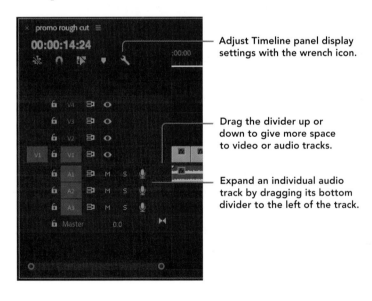

Figure 1.58 Use timeline controls to adjust the visibility of audio tracks and options.

Displaying the audio meters

If the audio meters are not visible, choose Window > Audio Meters (**Figure 1.59**).

As you watch audio levels for your sequence, you typically want to see audio levels that are high (between –6dB and 0dB) but never extending into the red area at the top. If your audio levels reach the red zone, your viewers will hear unpleasant distortion. You also don't want audio levels to be too low, or your viewers will have to turn up their volume more than they do for other videos.

Adjusting track volume

When you have one audio track, all you have to do is make sure that one track never gets too loud or too soft. When you have several audio tracks, make sure that they are at reasonable levels relative to one another; for example, background music should be low enough for a voiceover to be heard clearly. Also, there may be times when the combined levels of multiple tracks peak at an excessive audio level, producing distortion. Those are some of the reasons you may need to change the audio level of a track or vary it over time.

Figure 1.59 The Audio Meters panel helps you monitor audio levels in a sequence.

There's a horizontal **rubber band** control (a line) across the center of each audio clip; you can drag it up and down to adjust volume for that clip (**Figure 1.60**). When the rubber band is level, the audio level is constant; you can vary the level over time.

Figure 1.60 Drag the rubber band vertically to adjust clip volume.

TIP

When dragging a rubber band on a track, you can drag beyond the top and bottom of the clip. That way, you have finer control over the levels by dragging over a larger area.

To vary the audio level over time, add **keyframes** and adjust them.

1 Use one of these techniques (**Figure 1.61**):

- Select the Pen tool ( ) and click the audio level rubber band to add keyframe points at the times where you want the level to change.

- Position the playhead at the time where you want to add an audio keyframe, and then click the keyframe button for that track. If you don't see the keyframe controls, increase the height of the audio track until they appear.
- If the Selection tool is active, Ctrl-click (Windows) or Command-click (macOS) a track's audio level rubber band at the time where you want to add a keyframe.

In the example in video 1.15, four keyframes are added so that the audio level of the middle segment can be lowered without changing the rest of the track.

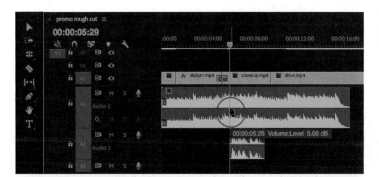

With the Pen tool, click the rubber band.

Click the keyframe button.

Ctrl/Command-click the Selection tool on the rubber band.

Figure 1.61 There's more than one way to add a keyframe to a rubber band.

2   With the Selection tool, drag the points you added up or down to shape the audio level over time (**Figure 1.62**). You can also drag a segment.

Figure 1.62 Dragging points added to the rubber band line to vary clip volume over time

★ ACA Objective 4.2

▶ **Video 1.16**
Adding a Simple Title

Adding a Simple Title

You'll probably want to add text to your videos, in the form of a title at the beginning and credits at the end, and possibly some clarifying overlay text along the way. In video editing, the text added to video is called *titles*. Premiere Pro CC 2018 includes redesigned titling tools that are easier to use than the titling tools in earlier versions.

Titles are typically added on a video track higher than the video clips, such as track Video 2, so that full-screen video clips don't obscure the added text. Check your Timeline panel to make sure there's enough room to work with the Video 2 track. If there isn't, allocate more vertical space to the video tracks. In the area containing controls just to the left of the track area (not in the track area), position the pointer over the double line dividing the video and audio sections of the timeline, and drag down to give more space to the video tracks.

In the project that Joe demonstrates in video 1.16, he adds a title to reinforce the "We get the dirty jobs done right" voiceover. To add a title in the same way:

1   In the Timeline panel, position the playhead at the time when you want the title to appear (**Figure 1.63**). In video 1.16, the playhead is positioned where the voiceover on the Audio 2 track begins.

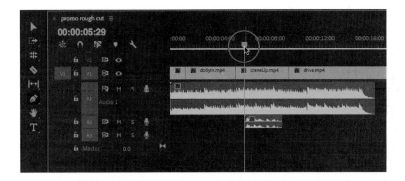

Figure 1.63 Position the playhead at the time the title should appear.

2. In the Tools panel, select the Type tool (T).
3. In the Project panel, click the Type tool where you want to start typing (**Figure 1.64**). (You can also drag to define an area for multiple lines of text.)

Figure 1.64 Click the Type tool on the video frame in the Program panel.

4 Type the text for the title, and then exit text editing mode by doing one of the following:

- In the Tools panel, select the Selection tool. A bounding box with handles appears around the title (**Figure 1.65**); you can now use the Selection tool to resize the title by dragging a handle, or you can reposition the title by dragging it from the middle.
- Press the Esc key. This might be a better method if you want to stick with the Type tool to add another title object.

Figure 1.65 After you're done typing the text, exit text editing mode to return to video editing mode.

NOTE

If you're editing title attributes in the Essential Graphics panel and you can't see the title or the effects of your changes, make sure the playhead is within the time that the title appears in the timeline.

When positioning a title in the Program panel, take care to keep it within the **safe margins** (see "About Safe Margins") so that it won't be cut off on some screens.

As long as the title is selected in the Timeline, you can now edit the title in various ways, including:

- To change the title's type and object attributes, including font, type size, and drop shadow options, choose Window > Essential Graphics to open the Essential Graphics panel. Then click Edit, and adjust the options there (**Figure 1.66**).

- To edit the text, in the Project panel either double-click the text with the Selection tool or click the text with the Type tool so that it displays a text insertion point. Then edit the text as you would in other applications (such as highlighting text to replace it).
- To change the attributes of individual characters, first enter text editing mode and select the characters, and then change the attributes in the Essential Graphics panel. Keep in mind that some Essential Graphics options, such as alignment, affect entire paragraphs or the entire title.
- To adjust the duration of the title, drag either end of it in the timeline, just as you would with a video or audio clip.

NOTE

The font menu isn't labeled in the Essential Graphics panel, but it's the first drop-down list under the Text heading.

Figure 1.66 The Essential Graphics panel contains the type controls that you can use to format a title.

ABOUT SAFE MARGINS

You may see a double set of margin guides around the edges of the Title panel. These are called *safe margins* (**Figure 1.67**). The inner margin is called the safe title margin, and the outer margin is called the safe action margin. To show or hide them, click the wrench icon in the Program panel and choose Safe Margins.

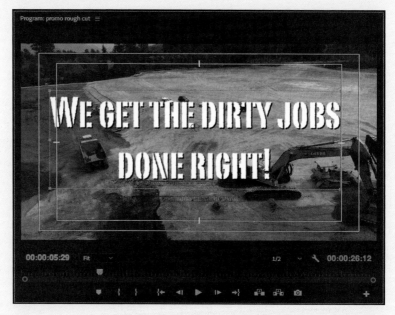

Figure 1.67 Safe action margin (outer) and safe title margin (inner) in the Title panel

Safe margins were more important back in the days when many televisions were set at the factory to enlarge the picture slightly to better fill the smaller screens at the time. This effect was called **overscan**, partly because enlarging would cut off the edges of the image. There was no standard for how much to overscan, so to make sure important content was not cut off on various televisions, the industry developed the concept of safe margins that were thought to be visible on most TVs. Videographers and video designers were instructed to keep all important content inside the safe action margin and keep all text inside the safe title margin.

continued from previous page

If you're creating a video program that may be distributed widely and viewed on a range of displays, including older televisions, it can be a good idea to respect the safe action and safe title margins. But the safe margins are much less necessary for video that will be shown on recent digital displays such as computer screens or high-definition televisions, because overscan is much less likely to be used on them.

Using Video Transitions and Effects

★ ACA Objective 4.6

▶ **Video 1.17** Using Video Transitions and Effects

After you've trimmed sequence clips to the point where you're probably done editing, you can start to enhance your sequence by applying video transitions and effects to clips. The main difference between the two is that you apply a transition only to the beginning or end of a clip, and a transition can affect the clips on both sides of an edit.

To add an effect to a clip:

1 Open the Effects panel (Window > Effects), and expand the Video Effects category (**Figure 1.68**).

Figure 1.68 Reveal the Video Effects category in the Effects panel.

2 Drag the effect you want, and drop it onto a clip in the Timeline panel (**Figure 1.69**). The clip now displays a colored fx badge.

Figure 1.69 Drag an effect onto a clip.

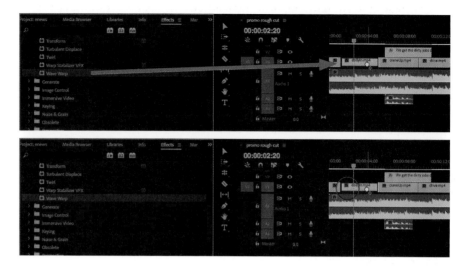

To add a transition to a sequence:

1 In the Effects panel, expand the Video Transitions category.
2 Drag the transition you want, and then in the Timeline panel, drop the transition at the edit point between two clips or onto the start or end of a clip (**Figure 1.70**).

Figure 1.70 Drag a transition and drop it between two clips.

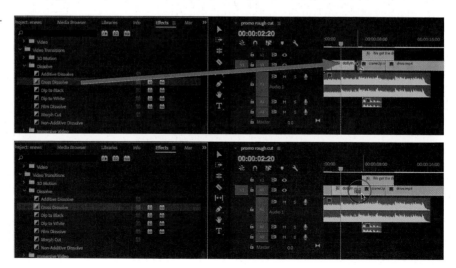

You can edit the duration of a transition directly in the Timeline panel by dragging either end with the Selection tool, just as you edit the In point or Out point of a video or audio clip. The transition will automatically maintain an equal amount of time before and after the edit, but if you want to adjust only one side of the transition, Shift-drag either end of the transition with the Selection tool.

You can make fine adjustments to effects and transitions in the Effect Controls panel:

1 Choose Window > Effect Controls to open the Effect Controls panel.

2 In the Timeline panel, select the clip containing the effect you want to edit, or select the transition you want to edit.

3 Use the options in the Effect Controls panel to edit the effect or transition (**Figure 1.71**).

If you've applied multiple effects to the clip, you'll see each effect listed with its settings in the Video Effects or Audio Effects sections of the Effect Controls panel.

> **TIP**
>
> *Just before you drop an effect or transition, pay attention to which parts of clips are highlighted under the pointer. For example, you'll usually want to make sure the clips on both sides of an edit are highlighted before you drop a transition between those clips.*

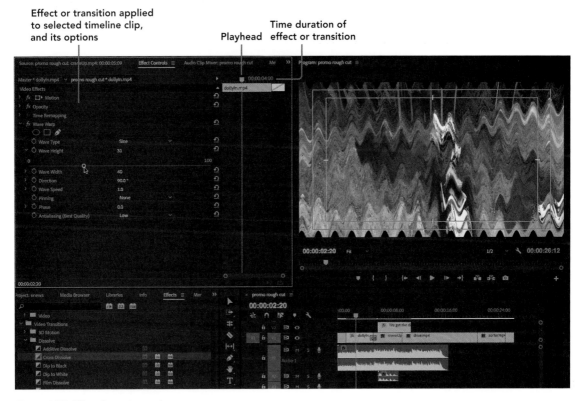

Figure 1.71 Effect Controls panel

Chapter 1 Introduction to Adobe Premiere Pro CC   **77**

Setting up a default transition

To save time, you don't have to drag and drop every transition you want to use. You can select clips on the timeline and choose Sequence > Apply Video Transition. Or to apply both the default video and audio transitions, choose Sequence > Apply Default Transitions To Selection. These commands both have keyboard shortcuts, which can be the fastest way to apply a transition. For Apply Video Transition, press Ctrl+D (Windows) or Command+D (macOS); for Apply Default Transitions To Selection, press Shift+D.

But what if you don't like the default transition? Choose your own. To change the default transition:

1. Open the Effects panel, expand the Video Transitions bin, and then expand the Dissolve bin. The current default transition is outlined.
2. Select the transition you want to use as your default transition.
3. In the Effects panel menu, choose Set Selected As Default Transition (**Figure 1.72**).

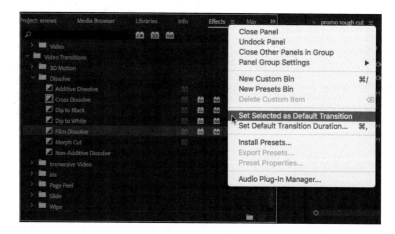

Figure 1.72 Cross Dissolve is the current default transition, but after applying Set Selected As Default Transition, Additive Dissolve will become the new default transition.

Applying an audio transition

Audio transitions work the same way as video transitions:

- You can apply an audio transition by dragging it from the Effects panel and dropping it in the timeline between two audio clips, or on one end of an audio clip (**Figure 1.73**).

- You can apply the default audio transition by selecting adjacent audio clips in the timeline and choosing Sequence > Apply Audio Transition or pressing Shift+Ctrl+D (Windows) or Shift+Command+D (macOS).
- You can change the default audio transition the same way you do for a video transition.
- When an audio transition is selected in the timeline, you can edit it more precisely in the Effect Controls panel.

Figure 1.73 To smooth the edit from one audio clip to another, drop an audio transition at the edit point between them.

USING TRANSITIONS WISELY

Premiere Pro comes with a long list of video transitions. Some of the fancy transitions are often overused by beginners, distracting from the story. As with any video tool, choose transition effects based on how well they serve the way you want to tell the story. The next time you watch TV or a movie, note which transitions are used for the type of program. You'll probably find that professional editors use transition effects sparingly, sticking to basic cuts and fades most of the time.

ENSURING ENOUGH EXTRA CLIP TIME FOR A TRANSITION

▶ Video 1.18
Ensuring Enough Extra Clip Time for a Transition

After you add a transition, you may find the transition not as long as you intended, or you might get an error message about "insufficient material." This can happen when at least one of the clips involved in the transition doesn't have enough extra material to make it work. For example, if you want a one-second transition centered between clips, the first clip must have an extra half-second of footage after its Out point, and the second clip must have an extra half-second before its In point. If that extra time wasn't captured in the original clip, you have to choose among these alternatives:

- Shorten or delete the transition
- Let the transition extend into the scene time
- Let Premiere Pro repeat end frames to create the extra time

In many cases those alternatives are not ideal. To avoid having insufficient time to create a transition, it's best to record more video than you need for a particular clip. Start recording a few seconds before the intended beginning, and stop recording a few seconds after the intended end. Many videographers don't stop the camera between takes, as long as there is enough free space on their media.

★ ACA Objective 5.1
★ ACA Objective 5.2

▶ Video 1.19
Exporting a Finished Video File

Exporting a Finished Video File

The sequence you play back in the Program panel is not yet a finished video; it's still a temporary combination of the clips and other media that Premiere Pro is holding together for you as you continue to edit it, and it doesn't yet exist outside of Premiere Pro. To get a copy of the project as a single video file, you must export video from the sequence.

Before you export a sequence, it's a good idea to play back the entire sequence in the Program panel so that you can spot any remaining problems and fix them. In video 1.19, that's how Joe realized he'd like to make one more edit to sync a cut with the music.

When you export, Premiere Pro not only has to assemble all of the components of a sequence into a single document, but it might also have to convert all the media in the sequence to a different format and compress data to keep the file size down.

Doing those tasks for every single frame will keep your computer very busy and can take a long time. Higher resolutions (such as 4K frames), heavy use of effects, and longer sequences can extend the exporting time further; in some cases, it takes several hours. You can cut down that time by using a computer with a faster CPU, more CPU cores, or a more powerful graphics card that's compatible with Mercury Playback Engine GPU Acceleration.

★ ACA Objective 5.1

You set up an export using the Export Settings dialog box. It includes multiple panels containing a large number of settings. There's no reason to let all those settings intimidate you, because you have an easy way out. As long as your intended output is a standard destination, you can set up most of the export settings in one step by simply choosing the preset for your destination.

▶ **Video 2.15** *Clean Up the Timeline and Export Your Project*

For this exercise, you'll export a sequence with settings that are consistent with the job requirements reviewed at the beginning of this chapter: a high-quality video file compressed for efficient loading over an online video service such as YouTube.

To export the sequence:

1. Make sure the sequence you want to export is either active in the timeline or selected in the Project panel or a bin.
2. Choose File > Export > Media (**Figure 1.74**).

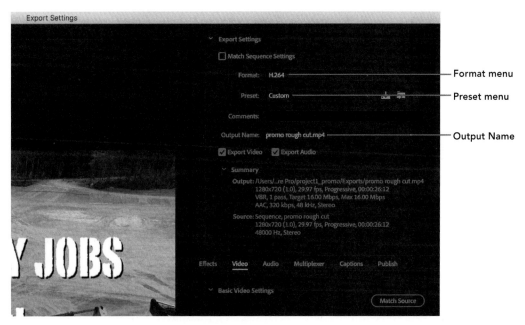

Figure 1.74 Set up an export in the Export Settings dialog box.

3. Click the Format pull-down menu and choose H.264.

   This step is important because the format determines which presets are available in the Preset list.

4. Click the Preset pull-down menu and choose YouTube 720p HD.

   That's the specification Brain Buffet is using for final output of the client video. This single step has now altered all export settings to match the selected preset.

5. Click the blue Output Name text, set the location (the Exports folder for this project) and filename (**enewsletter.mp4**) for the exported video (**Figure 1.75**), and click Save.

Figure 1.75 Always set the filename and location for an exported file.

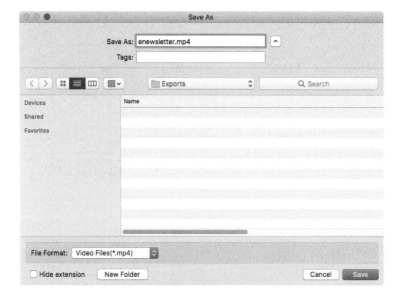

6. Click Metadata. In the Metadata Export dialog box (**Figure 1.76**), you can enter rights and licensing information, keywords that make it easier to find the video online, and so on. Enter any metadata that your project or client requires, and click OK.

If you expect to enter the same metadata for many projects, in the Metadata Export dialog box you can create Export Templates that contain that metadata. Then all you have to do is select a template to fill in the dialog box.

Figure 1.76 If your project requires entering metadata, add it in the Metadata Export dialog box.

TIP

You may be accustomed to being able to add metadata to photos at any time. For video, it's generally much easier to add metadata to a video during export than after it's been exported.

7. On your computer desktop, go to the Exports folder where you saved the video in step 5. Double-click enewsletter.mp4 to open it in your computer's video player, and check to make sure it plays back as you expected.

Exploring the Export Settings dialog box

There were many options that you didn't have to adjust individually because the preset you chose took care of them. But it's still good to have a general understanding of the different parts of the Export Settings dialog box, so let's take a brief tour of them.

The top section of the Export Settings dialog box contains the essentials: the format, preset, and filename and folder location.

In the middle of the Export Settings dialog box are six tabs of options. Again, these are automatically set up by the preset you choose, so in many cases you don't have to touch any of them. But if you do need to customize them, here's what each of the tabs covers:

- **Effects (Figure 1.77):** Sometimes you want to apply an effect only to an exported video, without applying it to the original sequence. One reason is when the sequence needs to meet specific technical requirements for delivery, such as video limiting or loudness normalization. Another reason would be to overlay a logo that you don't want to have on the original sequence.
- **Video (Figure 1.78):** If there's one tab you might adjust, it's the Video tab, because you can change the frame size and the bitrate. Lowering the bitrate is one tactic for decreasing the file size of an exported video; you can lower it until you notice decreased image quality.

Figure 1.77 You can use the Effects tab to modify the exported video without altering the original sequence.

Figure 1.78 In the Video tab, you can customize video export options.

- **Audio (Figure 1.79):** The Audio tab lets you configure details such as the audio format and bitrate.
- **Multiplexer (Figure 1.80):** You'll probably leave the Multiplexer tab as it was configured by the preset you chose, unless you understand MPEG multiplexing.

Figure 1.79 In the Audio tab, you can customize audio export options.

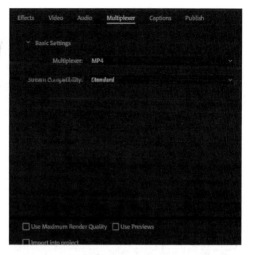

Figure 1.80 The Multiplexer tab contains MPEG multiplexing options.

- **Captions (Figure 1.81):** If you added closed captions to the sequence, in this tab you can decide how to export them. The sequence in this lesson doesn't use them.

- **Publish (Figure 1.82):** If you're exporting a sequence that's ready to be delivered straight to social media, you can set up your social media accounts in this tab, and the video will be uploaded after export.

Figure 1.81 You can adjust caption export options in the Captions tab.

Figure 1.82 To upload your video after it finishes rendering, configure the options for online services in the Publish tab.

If you do customize the settings, you can save them as your own preset by clicking the Save Preset button ( ).

If you want to know more about all of the options in the Export Settings dialog box, refer to the Export Settings Reference:

https://helpx.adobe.com/media-encoder/using/export-settings-reference.html

> ### ADOBE MEDIA ENCODER CC OR DIRECT EXPORT?
>
> There are two ways to export a sequence to a video file. You can export it from Premiere Pro directly to a video file as you do in this lesson, or in the Export Settings dialog box you can click the Queue button to have the video exported by Adobe Media Encoder CC, a separate program that comes with Premiere Pro. Using Media Encoder has several benefits:
>
> - Media Encoder can process exports in the background, so if you'd like to return to Premiere Pro to work on other sequences while your exports are processing, you can.
> - Media Encoder can queue multiple exports. Instead of exporting a sequence from Premiere Pro and waiting until it's done to export the next one, you can send a number of sequences from Premiere Pro to Media Encoder, where they will be queued up and automatically processed in turn.
> - If you need multiple versions of a sequence and the only difference between the versions are the export settings, you can export that sequence once from Premiere Pro, duplicate it in Media Encoder, and change the export settings for each duplicate. This will be much faster than exporting a sequence multiple times from Premiere Pro to Media Encoder, and it will result in less clutter than creating copies of a sequence.
>
> Because Media Encoder is a standalone video converter, you can even use it on its own to convert video files from their current format to another, a process called *transcoding*. For example, you can drag multiple video files into the Media Encoder queue and apply an export settings preset that converts them all for YouTube.

Challenge

You've taken a look at getting media ready for production, and you've also gone on a grand tour of the general Adobe Premiere Pro CC user interface. Now it's time for you to build something yourself.

Create a short promotional video for your school, your club, or a local business or nonprofit organization. Make it about 15–30 seconds long. Combine a collection of video clips that highlight your subject with titles, music, and a voiceover.

▶ *Video 1.20*
E-newsletter
Challenge

If you need background music, you can search the web for music licensed under Creative Commons. There are several types of Creative Commons licenses, so make sure you agree to the details of the license for the music you use. A good source of music recommended by Joe Dockery (author of the videos that accompany this book) is Incompetech:

www.incompetech.com/music/royalty-free/collections.php

After you complete the project, share it with your client and publish it online.

Remember Joe Dockery's Keys to Success from the video:

- Keep it short.
- Plan. Sit down with the client first so that you agree on the specifics of what the video should do.
- Shoot good-quality video. Keep the camera steady, and control the quality of light.
- Pay attention to file management.
- Use music licensed under Creative Commons if you're on a tight budget.
- Share your creation with the world!

Conclusion

Congratulations! You should now have a solid understanding of the overall video editing workflow in Premiere Pro. You learned about file organization and Premiere Pro panels and workspaces. You saw how to organize source media folders for a project, create a new project, and add files to a project. You also learned how to create a sequence; add video, audio, and titles to that sequence and edit them together; and export the sequence to a finished video file.

You learned quite a lot already, and this is only Chapter 1! In the rest of the book, you'll explore those subjects in more detail.

CHAPTER OBJECTIVES

Chapter Learning Objectives

- Edit in the Timeline panel.
- Compare lift and extract edits.
- Export a JPEG format image.
- Create L and J cuts.
- Create a lower-third title.
- Insert a graphic into a title.
- Create rolling credits.
- Work with B-roll.
- Apply speed changes and time remapping.
- Adjust volume.
- Stabilize shaky clips.
- Merge separate video and audio clips.
- Export your video.

Chapter ACA Objectives

For full descriptions of the objectives, see the table on pages 279–283.

DOMAIN 1.0
SETTING PROJECT REQUIREMENTS
1.1, 1.2, 1.3, 1.4

DOMAIN 2.0
UNDERSTANDING DIGITAL VIDEO
2.1, 2.2, 2.3, 2.4

DOMAIN 3.0
ORGANIZATION OF VIDEO PROJECTS
3.1

DOMAIN 4.0
CREATE AND MODIFY VISUAL ELEMENTS
4.2, 4.4, 4.5, 4.6, 4.7

DOMAIN 5.0
EXPORTING VIDEO WITH ADOBE PREMIERE PRO CC
5.1, 5.2

CHAPTER 2

Editing an Interview

Your second Adobe Premiere Pro CC project is a short, interview-based highlight video about a Brain Buffet employee. As in previous chapters, this is a hypothetical exercise created to help you explore some aspects of editing in Premiere Pro CC. You'll practice new video editing skills while building on the ones you learned in the previous project.

Preproduction

As you've learned, production starts only after the project requirements are clearly understood, so it's time to review those before you begin:

- **Client:** Brain Buffet Media Productions
- **Target audience:** Young professionals and students from 18 to 28 years old
- **Purpose:** The purpose of producing the employee highlight videos is to help customers connect with the people at Brain Buffet. Communicating what Brain Buffet employees are passionate about will help customers connect more strongly with the Brain Buffet brand.
- **Deliverable:** The client expects a 1-to-2-minute video featuring an interview with an employee that pairs high-quality video clips; music with a positive, upbeat feel; and a **lower-third** title featuring the company logo. To load quickly online, the video should be in H.264 YouTube 720p HD format.

★ ACA Objective 1.1

★ ACA Objective 1.2

▶ Video 2.1
Snowboarding Highlight Video

Listing available media files

In this project, some media has already been acquired for the project. What do you have to work with?

- Interview clip
- Snowboarding shots
- Brain Buffet logo
- Background music—2 minutes of upbeat stock music

That set of media is sufficient to complete the job, so you don't need to acquire any more media. Editing can begin.

★ ACA Objective 2.1

★ ACA Objective 2.4

▶ **Video 2.2** *Organizing the Media Files*

▶ **Video 2.3** *Setting Up the Project*

Setting Up the Interview Project

You'll start the editing stage of production by practicing the project setup techniques you learned earlier in the book.

Organize project files

On your desktop, unzip the project files in the project2_snowboarding.zip file and organize them into folders, using the same techniques and folder structure you used in the previous project.

Create a new project

Now you'll set up the interview project.

1. Start Premiere Pro CC, and when the Start screen appears, click New Project. The New Project dialog box opens.
2. Name this project **interview**.
3. In the New Project dialog box, click Browse, navigate to the project2_snowboarding folder, then navigate to the Project subfolder, and save this project there (**Figure 2.1**).

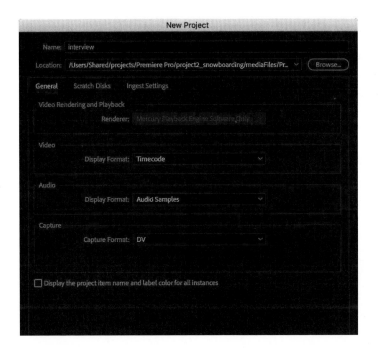

Figure 2.1 Choosing Project Settings in the New Project dialog box

The folder you applied becomes the default folder (Same as Project) for all of the locations in the Scratch Disks tab, and you can leave them as is for this project. You also don't need to change any settings for the Ingest Settings tab for this project.

Video 2.3 discusses the scratch disks and proxy workflows. To review how to decide where to assign scratch disks, see "Configuring the Scratch Disks tab" on page 21. Proxy workflows are covered in Chapter 3.

Importing with the Media Browser

With the project created, you can start bringing in the media you'll use. So far you've done this using the File > Import command or its shortcuts. But there's another way to import that has advantages in some situations: the Media Browser.

▶ *Video 2.4*
Importing with the Media Browser

Without leaving Premiere Pro CC, you can use the Media Browser panel to browse your computer and all of the volumes connected to it, including network volumes and media cards.

First, click the Media Browser panel to make it active. If you can't see it or its panel tab, choose Window > Media Browser.

Next, set up the Media Browser for efficient importing:

1 Position the pointer inside the Media Browser panel, without clicking.
2 Press the tilde key (`) to maximize the Media Browser and the other panels in its group.

The Media Browser (**Figure 2.2**) is organized in a fashion similar to file browsers you may have used on Windows or macOS. Use the panel on the left to select a storage volume and explore folders by expanding them. When you select a volume or folder, on the right you'll see the folders and files inside the selection.

Figure 2.2 Media Browser panel

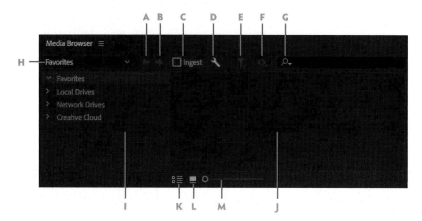

Always import only from a location that will be available at all times when you edit your project. For example, if you import from a camera card and then remove the card, the media is no longer available. Always copy media to your project's media folder before importing it; Media Browser ingest settings can copy for you.

A Back
B Forward
C Ingest (select to enable)
D Project settings, including ingest settings
E File type filter
F Directory viewers
G Search
H Recent locations
I Locations
J Subfolders and files
K List view
L Icon view
M Thumbnail size

Now import some media:

1 In the panel on the left, as needed, click the arrows to expand the volume and folders until you locate and select the project2_snowboarding folder, which contains the media you'll import for this project.

If the left panel is too narrow to show the entire paths and names of folders and files you're viewing, position the pointer over the divider to widen the left panel.

2. As you view the contents of the project2_snowboarding folder, hover the pointer horizontally across video clips to see how you can hover-scrub through those clip frames.

3. Select the items you want to import. For the snowboarding project, you can select all of the folders in the mediaFiles folder except for the Project folder (you don't need to import the project itself).

4. Do one of the following (**Figure 2.3**):
   - Choose File > Import From Media Browser.
   - Right-click (Windows) or Control-click (macOS) the selection and choose Import From Media Browser.
   - If the Project panel or a bin is visible, drag the selection from the Media Browser to the Project panel or bin.

> **NOTE**
> The term "bin" comes from the time when video editors used actual bins (containers) to organize physical videotapes for editing.

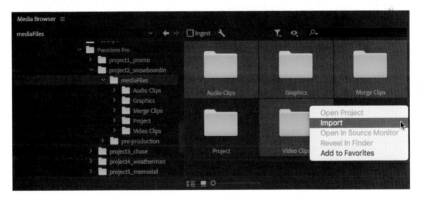

Figure 2.3 Importing with the Media Browser by right-clicking selected items

> **NOTE**
> The keyboard shortcut for the Import From Media Browser command is Alt+Ctrl+I (Windows) or Option-Command-I (macOS).

5. With the pointer inside the Media Browser, press the tilde (~) key to restore the Media Browser panel group to its normal size and position in the workspace.

6. Switch to the Project panel by clicking its tab, or press Shift+1. The three folders of media you imported are now available in the Project panel.

WHY USE THE MEDIA BROWSER?

It's important to understand that the Media Browser is not just another form of the Import dialog box. The Media Browser has several real advantages over the Import dialog box:

- The Media Browser is a more visual and interactive way to preview than in the Import dialog box. For example, in the Media Browser you can hover-scrub a clip to preview its frames.
- If you captured video in Advanced Video Coding High Definition (AVCHD) format, it will be easier to import using the Media Browser. Media captured in AVCHD appears as a folder containing many more folders and files, which you would have to reassemble in Premiere Pro CC. But in the Media Browser, you simply select the top-level folder of the captured media, and Premiere Pro CC assembles them into easy-to-handle video clips.
- If you want to apply Premiere Pro CC ingest settings, you don't have to wait until after the media is imported. Select the Ingest option at the top of the Media Browser, but also click the wrench icon to make sure the current ingest settings are what you want.
- You can click the Ingest Settings wrench icon to configure how ingest happens as you import.
- If you captured an event that the camera recorded as multiple video files (due to a limitation such as video file size or recording length), you can select multiple media items in the Media Browser, and when you import from there, Premiere Pro CC will create a single media item from the multiple clips you selected.
- If you're browsing media on a camera card, you can apply ingest settings that copy the files to your media drive so that they will still be available after you remove the camera card. You can do that instead of manually copying the files to your media drive before importing.

When using the Media Browser, remember that until you actually import an item, you are only browsing media on cards and drives. The true list of media imported into your project is in the Project panel, not the Media Browser.

Creating the Interview Sequence

★ ACA Objective 2.1

▶ **Video 2.5** *Creating a New Sequence*

In "Creating a New Sequence" in Chapter 1, you created a sequence from a clip so that the sequence settings would be based on the clip's settings. This time you'll do it a different way: you'll create a sequence from scratch, which means you'll have to specify sequence settings yourself. This is not as easy as creating a sequence from a clip, but you might need to do it this way when your production requires that the sequence have settings that are independent of clips from the cameras used in the production.

To set up a sequence from scratch, you'll use the New Sequence dialog box, which contains four tabs:

- **Sequence Presets.** Like other presets, Sequence Presets can save you a lot of time because they represent a wide range of industry formats and camera types. Applying one of these presets configures the options in the Settings tab for you.
- **Settings.** In this tab you specify video, audio, and video preview options for the sequence. They should be well matched to your production requirements as well as the hardware that you are using to run Premiere Pro CC. If you can select a preset in the Sequence Presets tab that is well matched to your production requirements and hardware, you may not have to change anything in this tab.
- **Tracks.** It isn't necessary to change anything in the Tracks tab, because you can easily add tracks to the sequence in the Timeline panel. If you already know that your sequence will require a certain number of video and audio tracks, it can be convenient to set them up here so they're all ready for you from the beginning.
- **VR Video.** If your sequence is intended for editing 360 video, you can prepare the sequence for that purpose in this tab. For this sequence, the VR Video tab isn't used.

To create and configure an empty sequence:

1. In the Project panel, create a Sequences bin in any of the ways you have learned.

    This step is not required, but when you work in a project containing many sequences, it's good to organize sequences into their own folder.

2. Open the Sequences bin you created.

3 Create a new empty sequence—not one based on a clip, but by doing one of the following:

- Choose File > New > Sequence.
- In the Project panel, click the New Item button, and choose Sequence.
- Right-click an empty area in the Project panel, and choose New Item > Sequence.

4 Click the Sequence Presets tab (**Figure 2.4**), and select a preset based on one of the following:

- The preset that's the closest to the type of camera you used to capture the sequence's video
- The preset that represents the postproduction requirements of your project; for example, if specific settings were provided to you by a postproduction supervisor

If you're following Video 2.5, expand the DV-**NTSC** folder and select Standard 48kHz. This does not match the clips that were shot, but it sets up the next section where you learn how to correct this.

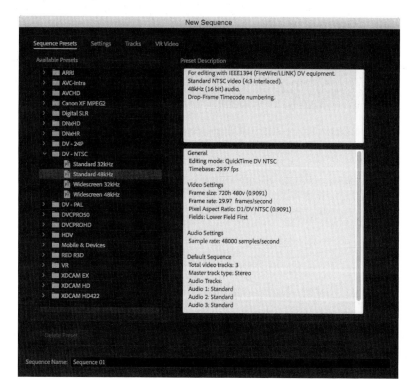

Figure 2.4 Choosing a sequence preset sets up the technical specifications of a new sequence.

5. Click the Settings tab (**Figure 2.5**), verify that the settings are appropriate for the sequence, and adjust any settings that need to be different from the selected preset.

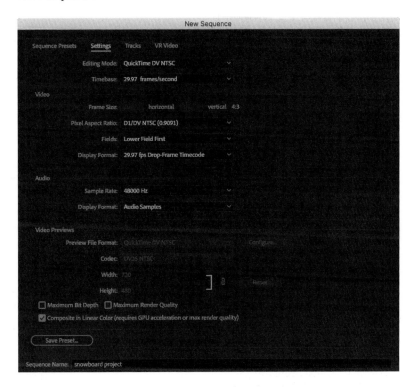

Figure 2.5 The Settings tab displays the video and audio specifications of the sequence.

6. If needed, repeat step 4 for the Tracks and VR Video tabs; you don't have to do anything with these two tabs if you're following Video 2.5.
7. For Sequence Name at the bottom of the New Sequence dialog box, enter a name. If you're following along with video 2.5, name the sequence **snowboard project**.
8. Click OK. The new sequence appears in the Project panel, the Program panel, and the Timeline panel.

Keep in mind that the sequence you use to edit does not necessarily have to match either the media you put in it or a specific output. You might record video with various 4K and 2K cameras, edit them together in a 4K sequence with a 5.1 surround audio mix, and use specific export settings to downmix the sequence to 2K video with a stereo mix.

But if you're working with just one camera on a simple production, like when you're just starting out, it's usually better not to go to all that trouble; just base a sequence on a clip from the one camera you have. You'll probably be specifying different settings at export anyway, starting with the format of the clips from your camera and ending by exporting to the format you need to upload the video online.

Handling a clip mismatch warning

When you import a clip into a sequence that uses different settings, the Clip Mismatch Warning alert may appear (**Figure 2.6**). Video 2.5 demonstrates an example of this, where Joe creates a sequence specified to be SD (standard definition) resolution and a 4:3 aspect ratio, but he tries to add an HD clip with a 16:9 aspect ratio.

Figure 2.6 Clip Mismatch Warning alert

If you are certain that the current sequence settings are correct, then all clips you add should adapt to them, and you should click Keep Existing Settings, the default button.

If you decide that the sequence settings should change to match the clip you're adding, click Change Sequence Settings.

To try this, do the following:

1. In the Project bin, open the Video Clips bin and drag the clip interview.mp4 to the Timeline panel, at the beginning of track V1.

    If the DV-NTSC, Standard 48Hz preset was applied in the New Sequence dialog box, the Clip Mismatch Warning alert appears because the clip differs from the sequence in several ways, including frame size and frame aspect ratio. In this case, incorrect sequence settings were applied when the sequence was created, and it would be better to have the sequence match the clip settings.

2. In the Clip Mismatch Warning dialog box, click Change Sequence Settings. The sequence settings now match the clip.

3. Save the project.

If the clip and sequence settings are quite different, the sequence may appear very different after you make your choice. In the example in Video 2.5, clicking Change Sequence Settings widens the sequence aspect ratio to match the incoming clip; if Keep Existing Settings was clicked, then the wide incoming clip would be cropped to fit the narrower aspect ratio and smaller frame size of the sequence's original settings.

To see how the sequence settings change after you choose Change Sequence Settings, choose Sequence > Sequence Settings. The resulting dialog box is identical to the Settings tab of the New Sequence dialog box.

When in doubt, build the sequence around the settings of your camera clips. This is a safer choice than making a wrong guess in the Sequence Settings dialog box.

Diving Deeper into the Workspace

Being able to customize your workspace can help you achieve maximum productivity in Premiere Pro CC, so Video 2.6 takes a second and more detailed look at the panels and workspaces in Chapter 1.

Reviewing workspace customization

★ *ACA Objective 2.2*

▶ *Video 2.6 Diving Deeper into the Workspace*

Take this opportunity to review and practice what you learned about Premiere Pro CC workspaces in Chapter 1. Also, explore the Workspaces panel and the Edit Workspaces dialog box.

At this point, you should know how to do the following:

- Open a panel that is not currently visible
- Resize a panel
- Rearrange panels by docking
- Rearrange panels by grouping
- Restore the last saved version of a workspace
- Choose a different workspace

Experiment and practice selecting, modifying, saving, and restoring workspaces. Remember how to use drop zones to control where a dragged panel will fall.

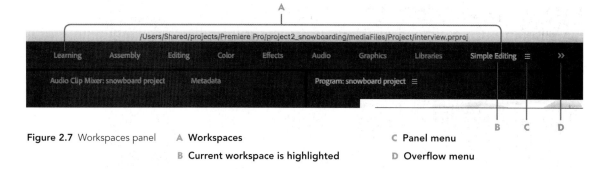

Figure 2.7 Workspaces panel
- A **Workspaces**
- B **Current workspace is highlighted**
- C **Panel menu**
- D **Overflow menu**

If you have not yet worked with the Workspaces panel (**Figure 2.7**), take a moment to explore it. The Workspaces panel is a convenient alternative to managing workspaces with the Window menu, because the Workspace panel can be visible at all times. To use the Workspaces panel:

1 If the Workspaces panel isn't visible, choose Window > Workspaces.
2 To switch to another workspace, click the name of the workspace.

If the list of workspaces is too long for the width of the Workspaces bar, click the overflow menu to choose a workspace name that isn't visible on the bar.

To customize the Workspaces bar, use the Edit Workspaces dialog box.

1 To open the Edit Workspaces dialog box, do one of the following:
   - Click the Overflow menu in the Workspaces panel, and choose Edit Workspaces.
   - Click the panel menu next to the name of the current workspace in the Workspaces panel and choose Edit Workspaces.
   - Choose Window > Workspaces > Edit Workspaces.
2 In the Edit Workspaces dialog box (**Figure 2.8**), drag to organize the names of workspaces in the list:
   - To change the order of the workspaces, drag them up or down in the list. For example, you might drag your favorite workspace to the top of the list so that it appears first in the Workspaces bar.
   - To make a workspace appear on the Workspaces bar, drag it into the Bar group.
   - To make a workspace appear in the overflow menu, drag it into the Overflow Menu group.

- To hide a workspace from both the Workspaces bar and the Overflow menu, drag it into the Do Not Show group.

3   Click OK.

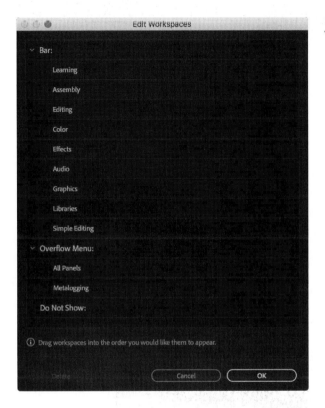

Figure 2.8 The Edit Workspaces dialog box

TIP

Can't see all of the workspaces at once in the Edit Workspaces dialog box? Instead of scrolling, you can make the dialog box bigger by dragging any corner or edge.

Taking a closer look at the Project panel

The Project panel is more than just a grid of previews for the media you've imported. There's a lot of information in there that you can use to help you efficiently identify and locate your project media.

▶ *Video 2.7* A Closer Look at the Project Panel

To explore the Project panel as demonstrated in Video 2.7:

1   If the Interview project you created earlier in this chapter isn't open, open it.
2   Open the Video Clips bin.
3   Switch to the Assembly workspace.

Figure 2.9 The Project panel with default thumbnails, larger thumbnails, and while sorting

In the Assembly workspace, the Project panel takes up the left half of the workspace. This arrangement is convenient when you're spending a lot of time organizing media in the Project panel. The Assembly workspace is also useful when you're putting together a rough cut; you aren't doing detailed refinements, so all you need are the Project, Program, and Timeline panels.

To customize the Project panel (**Figure 2.9**):

1 Drag the slider at the bottom of the Project panel to adjust the size of the media thumbnail previews.

You may want larger thumbnails to see them more clearly, or for more precision when hover-scrubbing. You may want smaller thumbnails to be able to see more of them in the Project window.

2 Click the Sort button at the bottom of the Project panel, and choose a criterion to sort by.

For example, you might want to sort the list by Video Usage to see which clips have already been used in a sequence. If the currently selected sort setting is User Order, that means the clips were manually arranged by dragging them.

Exploring List view in the Project panel

The thumbnail grid in the Project panel gives you a nice visual overview of the media in a project, but sometimes you might be more interested in seeing other types of information about the media. That's why the Project panel also offers List view. If you've used List view in the windows on your Windows or Mac desktop, you already know how this works: items are presented as a compact text list, with columns of metadata that you can use to sort the list.

To try out List view (**Figure 2.10**):

1. Click the List View icon at the bottom of the Project panel.

    When you'd rather see the metadata about a media item instead of a visual preview, List view can be quite informative.

    In List view, items are sorted by clicking a column header, similar to the way List view works in a Windows or Mac desktop window.

Figure 2.10 The Project panel in List view

2. Drag the horizontal scroll bar at the bottom of the Project panel to see the columns that are out of view to the right.

    Notice that some List view columns have gray values and other columns have blue values. The difference is that you can edit the blue values. For example, if you'd rather edit the In point and Out point of clips numerically instead of visually, click the blue text under Video In Point or Video Out Point to edit the time values by typing.

    Some List view fields are blank. You can enter text into many of these types of fields. For example, if you're logging video you just recorded, as you view a clip you can make a quick note of what it is by typing in the Description field for that clip.

To make List view easier to use, you can add or remove columns for the metadata you're interested in. You can do that using the Metadata Display dialog box.

To edit the List view metadata display:

1. Do one of the following:
   - Click the Project panel menu and choose Metadata Display.
   - Right-click any column heading and choose Metadata Display.
2. In the Metadata Display dialog box that appears (**Figure 2.11**), make sure any metadata properties you want to display in List view are selected, and then deselect any metadata properties you don't want to display. If you can't find the property you want to edit, do either of the following as needed:
   - Click the expansion arrow next to a metadata group to view the metadata items inside it.
   - Enter text in the search field at the top of the List view; the list will change to show only the items that match what you typed.

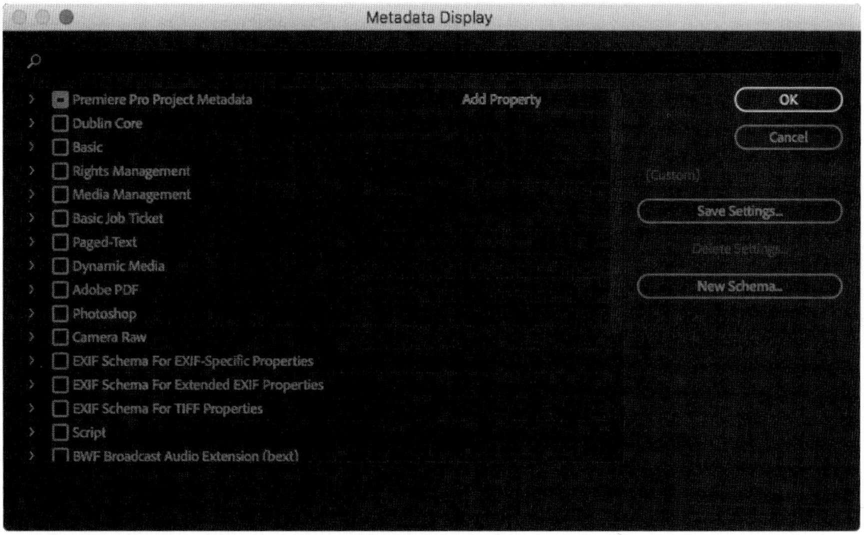

Figure 2.11 Use the Metadata Display dialog box to customize what you see in the Project window List view.

3. Click OK.

As you learn how you like to work in Premiere Pro CC, think about how you might want to adjust panel sizes, view modes, and positions for different display sizes (like your desktop and laptop). Remember that you can create easily switchable workspaces for different displays or purposes.

104   Learn Adobe Premiere Pro CC for Video Communication

Making Quick Fixes to Audio

★ ACA Objective 4.4
★ ACA Objective 4.7

▶ **Video 2.8** Audio Sweetening

Audio is just as important as video, especially in an interview. Just as you compose the elements of a video frame to work well together, you should be mindful of how well the different sounds in your audio track work together. For example, if one sound should be dominant, such as a person speaking, you probably want to de-emphasize other sounds such as ambient noise or background music by fading down their volume levels.

You haven't started editing the video for this interview yet, and part of the reason is that the key element in a video is the dialogue, which is in the audio. Therefore, it's better to start by editing to let the interview audio set the pace and then edit the video.

Using a subtractive workflow

In the promo project in Chapter 1, you trimmed and added a number of short clips to the timeline, where you assembled them. This chapter's interview project is built around one long interview clip, so it will be more efficient to use a different approach to editing it. You'll start with the long interview.mp4 clip you added to the timeline earlier, and then you'll cut out the parts you don't want.

Compare this subtractive style of editing to the additive style you used in Chapter 1:

- In additive editing, you set In and Out points for a clip in the Source panel, and then you add the resulting segment to the sequence, represented in the Timeline and Program panels. When you did this, you used the Insert and Overwrite options for adding clips to a sequence.
- In subtractive editing, you first add an entire clip to the sequence, and set In and Out points in the Timeline or Program panel to mark segments you want to delete. When you do this, you'll remove segments from the clip.

Using a subtractive workflow for the interview project simplifies some production tasks. For example, it's more efficient to make audio adjustments before you cut it up, while it's still all one clip. If you chop up the audio first and then adjust the sound of one of the resulting segments, you'll have to apply the same adjustment to the other segments, and that's more work.

Setting up a workspace for audio

Before editing the audio, it's good to review some best practices for preparing your audio workspace:

- Make sure the Project, Effects, Timeline, and Audio Meters panels are available. A shortcut is to simply choose a workspace where all of those panels are clearly visible. The Simple Editing workspace you created earlier shows all of the necessary panels, and so do the Editing and Effects workspaces.
- It may help to make room for audio editing in the Timeline panel. You can do this by vertically dragging the thick separator between the video tracks and audio tracks in the panel (**Figure 2.12**). Drag up to give more space to audio.

Figure 2.12 Dragging the separator between video and audio tracks

Applying an audio effect

The clip in the sequence has a problem, demonstrated in Video 2.8. All of the audio is in the left channel. This is not unusual, because an audio recorder is often configured to route each monophonic microphone to its own channel. But after being added to a video sequence, that microphone's audio needs to play evenly balanced across both the left and right channels. There's an easy way to do this in one step—it's an effect called Fill Right with Left.

You find both video and audio effects in the Effects panel. There's a long list of effects in the Effects panel, but fortunately you've got a shortcut—you can search the list:

1. In the Timeline panel, play back the sequence, which currently includes the interview.mp4 clip with signal in only one channel. You will hear audio in the left channel but not the right channel, and you can visually confirm this in the Audio Meters panel.

106 Learn Adobe Premiere Pro CC for Video Communication

2. Expand the view of the audio track so you can see the audio waveform in only one of its channels (**Figure 2.13**).

   You can expand a track by vertically dragging the horizontal divider line between tracks, by using the scroll wheel on a mouse, or by using the vertical scrolling gesture on a trackpad.

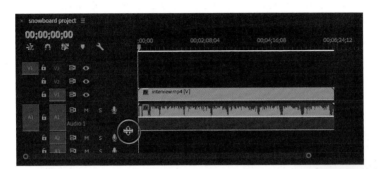

Figure 2.13 Dragging the audio track boundary down to make the track taller

3. In the Effects panel, click inside the search box at the top of the panel, and enter **fill** (**Figure 2.14**).

   You don't have to press Enter or Return; the search box returns results as you type.

4. Drag the effect Fill Right with Left onto the clip in the Timeline panel.

   The fx badge on the clip becomes purple to indicate that an effect is applied.

5. Play back the sequence. With the effect applied, the audio should now play through both the left and right channels.

6. Play back the audio while watching the Audio Meters panel. If the correct effect was added, the audio signal should now be indicated in both the left and right channel level meters, and you should be hearing the audio through both channels as well.

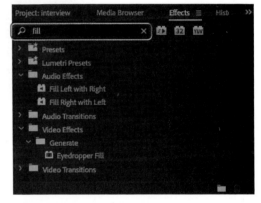

Figure 2.14 The Effects panel before (left) and after (right) entering **fill** in the search box

7. In the Effects panel, click the x in the search box to clear the search.

   Clearing the search is a good idea so that the next time you look for a different effect, the list doesn't just show you the results of the previous search.

Chapter 2 Editing an Interview 107

Using the Essential Sound panel

The audio for the interview clip contains some noise, so it would be great if that could be cleaned up. Fortunately, Premiere Pro CC has the Essential Sound panel. It provides ways to perform simple cleanups of audio so that you don't have to process the audio in a separate application. Now you'll apply Essential Sound to the audio clip:

1 If you weren't working in the Audio workspace before, switch to it now. The Essential Sound panel is one of the primary panels in this workspace.

Look around the Audio workspace and notice how it arranges and displays other panels (**Figure 2.15**). For example, the Audio Clip Mixer and Audio Track Mixer panels become the first two tabs in the panel group to the left of the Program panel.

Of course, you can also open the Essential Sound panel itself in any other workspace.

Figure 2.15 The Audio workspace, featuring the Essential Sound panel

A **Audio Clip Mixer (and Audio Track Mixer in grouped tab)**
B **Essential Sound panel**
C **Audio clip selected in Timeline panel**

2. Make sure the clip you want to repair is selected in the Timeline panel.

3. In the Essential Sound panel under Assign An Audio Type To the Selection, click the Dialogue button (**Figure 2.16**).

   When you assign an audio type, the Essential Sound panel displays options specific to that audio type—in this case, dialogue. The panel shows only the options relevant to the kind of audio that's selected, instead of overwhelming you with many more options that may not be useful.

   The Essential Sound panel simplifies audio editing in another way that you've seen in Premiere Pro CC before: it lets you choose a preset that applies settings as a useful starting point for the type of adjustment you need.

4. Choose Balanced Male Voice from the Preset menu. The Essential Sound panel changes the settings accordingly.

5. Click Repair to expand the Repair options, and then select Reduce Noise and Reduce Rumble (**Figure 2.17**).

   You can expand each of the rectangles in the Essential Sound panel, even though they don't display expansion triangles.

6. Play back the sequence. The background noise should be reduced.

7. Adjust the Reduce Noise and Reduce Rumble options to improve the adjustment until it sounds good to your ears. You can do this while the clip plays.

   Note that rumble is a form of low-frequency noise, so you may not hear it unless you're listening through headphones or full-range speakers.

   NOTE

   If you want to change the audio type assigned in the Essential Audio panel, click Clear Audio Type.

Figure 2.16 Clicking the Dialogue button in the Essential Sound panel reveals speech-specific options.

Figure 2.17 The Reduce Noise and Reduce Rumble options become available when you reveal the Repair options.

Chapter 2 Editing an Interview 109

Taking an approach to sound design

You did a little sound design work earlier in this book when you adjusted the audio levels for clips. Here are a few more tools and thoughts to keep in mind as you refine the sound design of your projects:

- Be sure to listen—not just to the sounds you know about, but to sounds that might be lurking in the mix and surfacing at times when they shouldn't.
- Listen to the relative levels of the different clips on a single track. To hear the audio for only one track, click the Solo button for that track. You've learned how to even out the levels of clips to match each other using the audio-level rubber bands on each clip.
- Listen for the relative levels of the various tracks in a sequence. You can adjust the audio level for each track by using the level knobs in the Audio Track Mixer panel.
- If you need to delete the audio track of a video clip or if you want to separately edit the durations of the video and audio tracks of a clip, select the clip and choose Clip > Unlink.

★ ACA Objective 1.4

★ ACA Objective 4.5

▶ *Video 2.9* Basic Color Correction

Making Quick Fixes to Color

If you look around in Premiere Pro CC, especially in the Video Effects panel, you'll find a wide variety of color controls. There are so many that it's understandable if you're intimidated by the sheer number of them. Many of the color controls are designed for editors who were trained to use traditional video color correction tools. But today there are ways to correct color that should be more familiar if you aren't already a trained colorist.

The approach to editing the video portion of the interview clip is the same as for the audio: correct the whole clip first, and cut it up later. You'll correct color with the easy-to-use Lumetri Color panel.

Using the Lumetri Color panel

The Lumetri Color panel presents a set of color controls that you may find easy to use, if you're familiar with the controls in photo editing software such as Adobe Lightroom or Adobe Camera Raw.

Figure 2.18 The Color workspace, featuring the Lumetri Color panel

A Lumetri Scopes panel: Color monitoring tool (not used in this exercise)
B Program panel: Preview color adjustments for selected clip
C Lumetri Color panel: Make color adjustments for selected clip
D Selected clip in the Timeline panel

To prepare for color correction, switch to the Color workspace (**Figure 2.18**).

Notice the panels that are brought forward in this workspace. The Lumetri Color panel is now along the right side, and Lumetri Scopes appears as the second tab in the top-left panel group. All audio panels are hidden except the Audio Meters panel.

CORRECTING WHITE BALANCE

The Lumetri Color panel follows the same basic design as the Essential Sound panel you just used. It contains a stack of expandable subpanels.

1. Click White Balance to reveal white balance options.

   As in photo editing applications, white balance controls the overall color balance of the image. **Temperature** adjusts white balance along an axis from cool (blue) to warm (orange), whereas Tint adjusts along an axis from green to magenta. Instead of making time-consuming manual adjustments, you can instantly adjust white balance by using the WB Selector (eyedropper).

2. Click the WB Selector eyedropper icon, and then click an area of the image that should be neutral gray or near-white. The Temperature and Tint values should change, and the image should appear more neutral (**Figure 2.19**).

Figure 2.19 Clicking the White Balance eyedropper on an area that should be neutral

It's best to click on an area that is not completely white. If an area is so bright that all color channels are at maximum luminance, the WB Selector won't find any color imbalance to correct.

After you make the image neutral, you can adjust Temperature and Tint to help achieve the look you want for the video.

ADJUSTING TONE AND SATURATION

You can use the Tone controls to change how brightness levels are distributed along the tonal range from black to white. For example, you can decide to make the image brighter in the shadows.

1. If the Tone options are hidden, click Tone to reveal them (**Figure 2.20**).
2. Adjust the tone options for good overall image brightness and to show the level of detail you want in the shadows and highlights:
   - **Exposure** affects overall brightness.
   - **Contrast** affects the overall difference between dark and light areas.

112 Learn Adobe Premiere Pro CC for Video Communication

- **Highlights** affects the visibility of details in the near-white areas of the image. In the interview clip, reducing Highlights darkens the bright areas in and near the sky, making the details in the clouds and ridge more visible. Reducing Highlights is also useful for getting more detail out of overexposed video.
- **Shadows** affects the visibility of details in the near-black areas of the image. In the interview clip, increasing Shadows lightens the black shirt, making shadow detail easier to see. Increasing Shadows is also useful for getting more detail out of under-exposed video.
- **Whites** affects the brightest areas of the image. For example, if an image is underexposed, you can increase Whites to more fully use the upper end of the tonal range.
- **Blacks** affects the darkest areas of the image. For example, if an image is overexposed, you can decrease Blacks to more fully use the lower end of the tonal range.

In general, set the Tone controls from the top down. For example, the Whites and Blacks values are easier to set after you've already adjusted Exposure and Contrast.

3 Adjust Saturation to the increase or decrease the overall color intensity.

Figure 2.20 Tone options expanded

> **TIP**
> If you're not sure what changes to make in the Tone settings, try clicking the Auto button.

PREVIEWING AND RESETTING CHANGES

Each group of Lumetri Color settings has a check box to the right of the settings group name. When a settings group check box is selected, the settings are applied to the selected clip. You can deselect the check box to see how the clip looks when that group of settings is not applied, so the check box is a quick way to do a before-and-after comparison.

The Tone settings group also has a Reset button, which changes all of the values to zero in case you want to start over.

ABOUT OTHER LUMETRI COLOR PANEL OPTIONS

Of the subpanels in the Lumetri Color panel, you'll probably use Basic Correction most of the time. Once you've corrected any fundamental issues with a clip, you can then use the other Lumetri subpanels to apply expressive adjustments that fit the visual intention of your production:

- **Creative.** The Look option is a quick way to apply a cinematic visual style, called a , to a clip. Some looks emulate film stocks; others create a mood.

The Adjustments settings let you do things such as sharpen detail or apply different tinting to the shadows and highlights of a clip. The Vibrance setting is like a safer version of Saturation because Vibrance tries to avoid oversaturation, especially in skin tones.

> **NOTE**
> Because Lumetri Color is an effect, you can also adjust its options using the Effect Controls panel.

- **Curves, Color Wheels,** and **HSL Secondary.** These color controls are familiar to trained video colorists. Because they're advanced controls, they're not covered in this book.

- **Vignette.** Vignetting is a technique that helps focus the eye on the center of a frame by darkening or lightening the edges. It's a time-tested technique that's been used in painting for hundreds of years, and it works for video too. It's best applied at a small, subtle amount.

ACCELERATING THE LUMETRI COLOR PANEL

If you use the Lumetri Color panel and it takes a few moments for the changes to appear on your computer, you may benefit from graphics acceleration. Choose File > Project Settings > General, and in the Renderer menu, choose an option that says GPU Acceleration if one is available. That should result in faster, smoother previewing of clips that have adjustments applied using Lumetri Color or any other GPU-accelerated effect.

If a GPU Acceleration option is not available (if the only option says Software Only), consider upgrading your computer with graphics hardware supported by the Mercury Playback Engine (see "Accelerating Performance with the Mercury Playback Engine" in Chapter 1).

★ ACA Objective 4.1
★ ACA Objective 4.3

Subtracting Unwanted Clip Segments

▶ **Video 2.10** Lift and Extract

Like any interview, ours has questions and answers. For a shorter, faster-moving interview, this interview will be edited so that you hear only the subject answering the interviewer's questions. The interviewer's questions will be cut out. A faster-paced interview is appropriate for an interview like this one, which will be intercut with action-oriented snowboard footage. Your job is to edit out the interviewer's questions, as well as any unwanted bits of noise such as coughing.

Marking part of a clip for removal from a sequence

You're done with the specialized editing work on the audio and video, and you're about to move on to clip editing. So the first thing you'll do is change the workspace. Then you'll mark and delete an unneeded segment of the long interview clip.

> **TIP**
> Looking at the audio waveform can help you spot the quiet passages you want to remove, such as in the interviewer questions in the interview.mp4 clip.

1. Switch to the Simple Editing workspace you created earlier.
2. In the Timeline panel, zoom out so you can see the entire interview clip.

   Remember that the backslash key (\) is a shortcut for fitting the entire sequence in the Timeline panel.
3. In the Timeline panel, move the playhead to the beginning of any part of the interview that you want to remove.

   A good place to start is the first frame of the interview clip so that you can cut out the preparation segment before Joe starts talking.
4. Mark the In point by doing one of the following:
   - In the Program panel, click the Mark In button (**Figure 2.21**).
   - Choose Markers > Mark In.
   - Press the I key.

Mark In Mark Out

Figure 2.21 The Program panel controller

The time rulers in the Program panel and Timeline panel both turn gray starting at the In point you set. The gray segment ends at the Out point, which is not yet set.

5. In the Timeline panel, move the playhead to the end of this part of the interview that you want to remove.

   If you set the In point at the beginning of the clip, a good place for the Out point is at 9 seconds, just before Joe starts talking.
6. Mark the Out point by doing one of the following:
   - In the Program panel, click the Mark Out button.
   - Choose Markers > Mark Out.
   - Press the O key.

Figure 2.22 Sequence In to Out point range marked in the Timeline panel

The gray segment on the time rulers now appears only between the In point and the Out point you set (**Figure 2.22**).

Notice the difference between the In point and Out point you applied here, compared to the In and Out points you applied for additive editing:

- In additive editing, you set clip In and Out points. They're attached to the clip and they appear in the Source panel, where you applied them. After you add the clip to the Timeline panel, the clip In and Out points become the starting and ending frames of the trimmed clip in the sequence.

- In subtractive editing, you set sequence In and Out points. They're attached to the clip instance in the Timeline because they're specific to that sequence. Sequence In and Out points don't appear in the Source panel.

It's important to understand the difference between clip and sequence In and Out points—for example, so that you can understand that a clip's In and Out points in the Timeline (which are sequence In and Out points) can be different than they are in the Program panel (which are clip In and Out points).

If you want to remove the In and Out points, choose Markers > Clear In And Out. You can also right-click (Windows) or Control-click (macOS) and choose Clear In And Out, or press Alt-X (Windows) or Option-X (macOS).

Removing a marked clip segment from a sequence

Once In and Out points are marked in a sequence, you're ready to modify the interview clip. You'll try this in two ways. First try simply removing the clip segment from the sequence (**Figure 2.23**):

> **TIP**
> If the current magnification makes it hard to see or navigate frames in the Timeline panel, magnify it. You can press the hyphen (-) and equal sign (=) keys to zoom out and in, respectively. It's easier to remember those keys by their respective uppercase characters, which look like a minus and a plus sign.

1. To remove the clip segment, do one of the following:
   - In the Program panel, click the **Lift** button.
   - Choose Sequence > Lift.
   - Press the semicolon (;) key.

> **TIP**
> To move the playhead by one frame, press the Left Arrow or Right Arrow key.

> **TIP**
> To move the playhead by multiple frames, type the plus sign (+) into the Playhead Position time display, type the number of frames you want to move, and press Enter or Return.

> **TIP**
> You can also use the Razor tool to slice a clip in two places, and then delete the resulting segment.

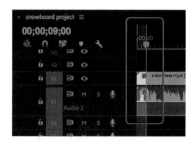 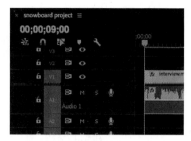 

Before edit | After using Lift: Gap results from removed segment | After using Extract: Following clips shifted to cover gap

Figure 2.23 Lift vs. Extract

The marked segment is removed, leaving a gap. Try this another way.

2. To remove the clip segment and automatically close the gap with a ripple edit, do one of the following:
   - In the Program panel, click the **Extract** button.
   - Choose Sequence > Extract.
   - Press the apostrophe (') key.

   The marked segment is removed, and the rest of the clip is shifted left to close the gap.

A clip doesn't have to be selected for Lift and Extract to affect it. Lift, Extract, and other clip modifications affect any tracks targeted in the Timeline panel. Remember that near the left side of the Timeline panel, the track targeting switches show which tracks are targeted. In the interview sequence, the clip video on track V1 and the clip audio on track A1 are edited by Lift and Extract because tracks V1 and A1 are currently targeted (**Figure 2.24**).

Figure 2.24 Track-targeting switches

> **TIP**
> *When you use Lift or Extract, the segment you lifted or extracted is available on the clipboard, so you can paste it at a different time or track.*

As with other clip edits you've done in a sequence, Lift and Extract don't permanently delete frames from the original clip. They simply hide all frames except the segment between each clip's own In point and Out point.

Chapter 2 Editing an Interview 117

TIP

When editing jump cuts, try to choose In points and Out points that have consistent framing.

Now, practice subtractive editing by using sequence In and Out points together with Lift or Extract to remove unwanted segments of the interview clip so that the only segments remaining are the most interesting ones where the subject is answering the interview questions. Try to edit down the roughly 6-minute interview clip to 1–2 minutes. That may seem like a lot to cut out, but if you were asked to prepare an excerpt from this clip for social media, it might need to be even shorter.

Removing gaps between clip instances

If you used Lift to remove some segments from the interview clip, you may now have gaps in the sequence. To remove the gaps, do any of the following in the Timeline panel:

- With the Selection tool, click to select a gap (**Figure 2.25**) and then choose Edit > Ripple Delete.
- With the Selection tool, click to select a gap and then press the Delete key.
- Right-click a gap, and choose Ripple Delete.

NOTE

You can also do a ripple delete of a selected clip by pressing Shift+Backspace (Windows) or Shift+Forward Delete (macOS); see the sidebar "Backspace/Delete and Forward Delete: Similar but Different Keys" in Chapter 1.

Figure 2.25 You can select and delete a gap the same way you would delete a clip.

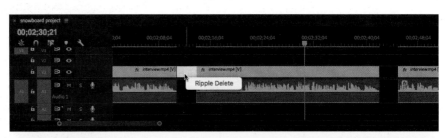

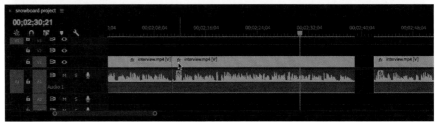

Getting Organized in the Timeline Panel

★ ACA Objective 4.1

▶ Video 2.11
Organization in the Timeline

As you work on a project, you might be building a project out of multiple sequences containing a large number of video, audio, and still image media files. Within a single sequence, you may be working with many clips on one track representing different talent or storylines, and you may be working with many tracks. As the project becomes more complex, disciplined organization becomes more critical to your success. Premiere Pro CC can help you organize your projects visually so that you can easily identify project components by looking.

When you use a clip more than once in a project, you create additional **instances** of the clip. Instances aren't complete copies of the original clip file; they're references to the clip in the Project window. You create instances when you drag the same clip into sequences multiple times or when you start with an entire clip in a sequence and cut it up into multiple segments.

Setting how clip organization affects clip instances

Before you start, there is one decision you should make. Do you want each instance of a clip to be able to have its own label and name, or do you want all instances of a clip to use the same colored label and name? This is a settings you can make at the project level.

To set how editing labels and names affect instances of a clip:

1 Choose File > Project Settings > General (**Figure 2.26**).

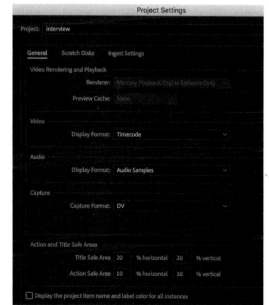

Figure 2.26 Display The Project Item Name And Label Color For All Instances preference

2 Do one of the following:
  - If you want all instances of a clip to change when you edit the label or name of any instance of that clip, select Display The Project Item Name And Label Color For All Instances.
  - If you want each instance of a clip to be able to have its own label or name, deselect Display The Project Item Name And Label Color For All Instances. For the interview project, this setting is deselected.

3 Click OK.

Applying colored labels to clips

One way to organize the Timeline panel is by applying colored labels to clips.

To change the colored label for a clip, do one of the following:

- With the Selection tool, click to select the clip, and then choose a label from Edit > Label.
- Right-click a gap, and then choose a label from the Label submenu.

TIP
You can customize the label names and colors in the Labels pane of the Preferences dialog box.

Changing clip names

Another way to identify clips is by their names in the Timeline panel.

To change the name of a clip:

1. Do one of the following:
   - With the Selection tool, click to select the clip, and then choose Clip > Rename.
   - Right-click a clip, and then choose Rename.
2. In the Rename Clip dialog box, enter a new clip name, and then click OK.

★ ACA Objective 1.5
★ ACA Objective 4.3

▶ **Video 2.12**
L and J Cuts

Applying L and J Cuts

So far, you've edited a clip's video and audio together: The video and audio parts of a clip start and end playing at the same time. But another common way of editing is to have slightly different edit points for the video and audio components of the same clip. The **L cut** and the **J cut** are two examples of this. For example, if you watch television shows or movies, it's common to start hearing the audio of the next clip before you see its video. This can create edits that feel more dynamic than when audio and video always use the same edit point.

The names L cut and J cut come from how they look in the timeline. In an L cut, the audio of a clip, shown lower in the timeline, extends further in the timeline than the clip's video on an upper track. In a J cut, the audio track of a clip starts before the video.

Adding B-roll clips

You can use L cuts and J cuts in this video, especially when editing **B-roll** (alternate or background footage) into the sequence. B-roll is used to help maintain the viewer's interest with engaging supplementary video or to provide informative visual background and context for the voiceover. In this example, the snowboarding clips both enhance your understanding of Joe's personality and give you an opportunity to relate to him through his snowboarding hobby. The B-roll also adds visual interest beyond simply watching him talk to the camera.

> **NOTE**
> Clips on higher tracks completely cover clips on lower tracks. If you want lower tracks to show through, you have to modify the clip on the upper track by lowering its opacity (in Effect Controls) or adding a mask to it.

Primary video is traditionally referred to as A-roll, and secondary footage is B-roll. The A and B terms came from tape-based video editing bays where you would load tapes into two source video tape decks marked A and B and assemble them into your sequence on your program video monitor. You can add B-roll clips into gaps left by the removal of parts of clips on the main video track, or to a higher track.

To add some B-roll to the sequence as an L cut:

1. Switch to the Assembly workspace.
2. In the Program panel, move the playhead to the time in the sequence when you'd like to insert some B-roll. For example, it might be over the end of the introduction, after the interview subject has talked for a few seconds. Make a note of how much time is available for the B-roll before the next edit.
3. In the Project panel, double-click a clip you'd like to use as B-roll. It opens in the Source window, which is grouped with the Project panel in this workspace.

    Be sure to avoid confusion as to which panel you're viewing above the timeline by paying attention to the panel names in the tabs.
4. In the Source panel, set In and Out points for the most interesting part of the B-roll clip. Make sure the duration fits within the time available for it in the sequence.

5 Position the pointer over the Drag Video Only icon in the Source panel, and then drag the icon to the intended time on track V1 in the Timeline panel (**Figure 2.27**).

Figure 2.27 Before (left) and after (right) adding the video of a B-roll clip to track V1

Figure 2.28 The L shape of an L cut, indicated in orange

> **TIP**
>
> To prevent audio on the main video track (usually V1) from being altered by edits on other tracks, in the Timeline panel you can click the lock icon () for the main video track.

6 Use editing tools as needed to refine the timing of the B-roll clip.

The Rolling Edit tool may be particularly useful because it lets you shift the edit point without changing anything else.

If you added the B-roll to the end of an interview clip, the video and audio of the interview clip now forms an L shape (**Figure 2.28**) around the B-roll clip as the audio continues under the video. That's why it's called an L cut.

If you had edited it so that the audio portion of the interview clip ends before the video does, then it would be a J cut.

A quick way to do an L cut or J cut is to Alt/Option-drag the end of a video or audio track. This changes the In point or Out point of just the video or audio portion of a track so that you can edit a clip's audio and video In and Out points independently.

How do you know which clips you've already used in a sequence? Earlier you saw that the Project panel can show video usage when set to List view. But the Icon view also shows the usage of video and audio. Each thumbnail displays badges in the bottom-left corner. When they're gray, they have not been used in a sequence. When they're blue, they have been used.

Playing a Clip Faster or Slower

★ ACA Objective 4.4

▶ *Video 2.13* Basic Speed Changes

The interview sequence is sports-oriented, and that's a great time to use slow motion on a B-roll action clip. This time you'll add a B-roll clip to a different video track and adjust its playback speed.

First, add the B-roll clip:

1 In the Project panel, double-click the B-roll video you want to use. Video 2.13 uses vid-nosetrick.mp4.

2 As you did earlier, set the In and Out points of the clip in the Source window.

3 Drag just the video portion of the clip to track V2; the exact time position isn't important yet. You're using the V2 track so that when its duration changes you don't have to be concerned about changing clips around it.

4 With the clip selected, choose Clip > Speed/Duration (or right-click the clip and choose Speed/Duration) (**Figure 2.29**).

5 In the Clip Speed/Duration dialog box, enter a speed much lower than 100%, such as **50**.

A speed lower than 100% slows motion, whereas a speed higher than 100% plays faster motion.

6 Click OK, and play the sequence through the B-roll clip to observe how the slow motion looks.

Notice that when you change the clip speed, the clip duration changes. That's because speed and duration are inversely related: the faster you play a clip, the shorter it can play.

If you want the clip to play for a specific length, you can change the duration instead and the speed will be adjusted to fill the new length.

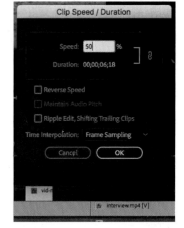

Figure 2.29 The Clip Speed/Duration dialog box, with your new B-roll clip selected in Timeline panel

If you think the B-roll clip should be faster or slower, edit the value in the Clip Speed/Duration dialog box.

7 When you like the slow-motion speed of the B-roll, drag it to the time in the sequence where it belongs. You can leave it on track V2 or drop it into track V1.

What if you want to visually adjust the playback speed of a clip or you're not sure how long of a duration you want to fill with it? You don't have to do the math or use trial-and-error with the Clip Speed/Duration dialog box. Instead, you can interactively adjust clip speed using the Rate Stretch tool.

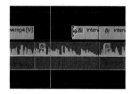

Figure 2.30 A J cut, indicated in orange

1. In the interview sequence, find a clip that doesn't yet have B-roll added to it.
2. Position the pointer near the beginning of the video of the clip.
3. Alt-drag (Windows) or Option-drag (macOS) the left end of the clip to the right about five seconds according to the tool tip so that the clip's video ends before the audio. This is an example of a J cut (**Figure 2.30**).

   There is now a gap to fill on track V2, and you'll fill it with B-roll.

4. In the Project window, double-click the clip you want to use as B-roll. Video 2.13 uses vid-jumpRed.mp4.
5. Set the In and Out points of the clip in the Source panel so that the duration is a second or two shorter than the gap that was created. You don't need to be exact because you'll adjust this soon.
6. Drag the video portion of the clip from the Source panel to the gap in the sequence.
7. Select the Rate Stretch tool ( ), which is grouped with the Ripple Edit tool ( ) in the Tools panel. Drag the ends of the B-roll clip until it completely fills the gap.

   Premiere Pro CC adjusts the clip duration by changing the clip speed, leaving the clip In and Out points unchanged (**Figure 2.31**). If you're curious how much the clip speed changed, select it and choose Clip > Speed/Duration.

Figure 2.31 Before (left) and after (right) using the Rate Stretch tool on a clip to fill a gap

Keep in mind that the amount of speed change depends on how much you stretch the clip. The longer the original duration of the B-roll clip, the less it would be stretched to fill the gap, and the less it would be slowed.

8. Play back the sequence to see the results, and make any necessary adjustments.

CREATING SMOOTH SLOW MOTION

Although you can apply a slow-motion speed change to any video clip, slow motion looks smoothest on clips recorded at high frame rates. For example, if you shot 30 fps video for a 30 fps sequence and you slow down a clip by 50%, that clip now plays at an effective frame rate of 15 fps, so the slow motion may not be completely smooth.

When you record clips that you know you'll want to play back in slow motion, set the camera to a higher frame rate if available. For example, if you plan to play back a clip at 50% speed in a 30 fps sequence, set the camera to record at 60 fps. When that clip is slowed by 50% its frame rate will become 30 fps, matching the sequence frame rate and appearing perfectly smooth.

When you slow down a clip so that its effective frame rate is lower than the sequence frame rate, Premiere Pro CC offers three ways to improve the quality of the slow motion:

- **Frame Sampling.** This is the simplest option. Frame Sampling reconciles the effective frame rate of the slowed clip with the sequence frame rate by repeating or leaving out frames as needed. For example, if the slowed clip has an effective frame rate of 15 fps in a 30 fps sequence, each clip frame will be repeated twice.
- **Frame Blending.** Instead of merely repeating frames, Frame Blending merges the previous and next clip frames into transitional frames. This requires somewhat more processing power than Frame Sampling.
- **Optical Flow.** This is the newest and most sophisticated option. Optical Flow analyzes the frame content and its pixel motion, and it renders new video frames. It works best when a moving subject is in front of a contrasting non-moving background and where there is no motion blur (such as in video shot at a high shutter speed). Optical Flow requires significantly more processing power than the other two options.

Playing a Sequence Smoothly

 *ACA Objective 3.1*

 *Video 2.14
Render*

Editing can be frustrating if playback isn't smooth. Simple one-track sequences might play back smoothly in the Source Monitor or Program Monitor, but playback may stutter or seem to be unresponsive if you use higher-resolution source footage (such as 4K media). The processing requirements of rendering a preview go up when you add effects or overlay tracks. All of these burdens are higher when you edit on a less powerful computer. Premiere Pro CC provides tools for letting you know how well your computer is performing, and it also provides ways for you to balance the trade-off between preview quality and smooth playback. For example, on a less powerful computer you can lower the preview quality or pre-render some parts of the timeline so that the computer can achieve smooth playback. On a more powerful computer, you can raise the preview quality as long as the computer can maintain smooth playback.

The render bars above the timeline (**Figure 2.32**) have different colors depending on the ability of Premiere Pro CC to keep up with the performance demands of your sequence.

- No color is okay. It means Premiere Pro CC is confident it can play back that media at full quality in real time without having to render a preview file for it.
- Green is okay. Premiere Pro CC has rendered a preview file for that segment and the preview file is up to date, so playback will be at full quality in real time.
- Yellow is usually okay. Premiere Pro CC thinks it can play back at full quality, but you might see occasional stuttering.
- Red means that Premiere Pro CC thinks that segment is complex enough that you're probably going to see a lot of stuttering and delays. For real-time playback, you'll want to choose one of the Sequence > Render commands.

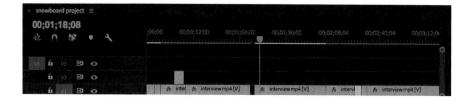

Figure 2.32 Colored render bars above the timeline

The same sequence might display different bar colors on another computer, because the colors are about the complexity of the sequence and media compared to the capabilities of the current computer. If Premiere Pro CC is having trouble playing back your sequence smoothly, there are several ways to address the problem:

- Adjust playback resolution in Premiere Pro CC. The Source Monitor and the Program Monitor both have a Select Playback Resolution menu. If you choose a lower resolution, you will lower the amount of work Premiere Pro CC has to do to render each frame, making it easier to achieve smooth playback.

- Render all parts of the timeline that are marked with a red bar by pressing Enter or Return. Note that render does not mean export; this is about pre-rendering previews of the timeline so that your computer doesn't have to try to render them during real-time playback. Of course, the disadvantages of pre-rendering are waiting for those previews to complete rendering and having to redo previews when you make edit the clips they represent.
- Use proxies. A proxy is a copy of a clip that plays back faster because its quality has been lowered. You'll learn more about proxies in Chapter 3.
- Upgrade your computer. The specific upgrades you need depend on where the bottleneck is. You may need to add more RAM, use faster storage, spread scratch files across more drives, or add a graphics card that's supported by the Mercury Playback Engine so that you can take advantage of GPU acceleration.

PREVIEWING TRANSITIONS

You might notice that rendering bars indicating lower performance show up more often over transitions than over a single clip. Why is that?

It isn't hard to play back a single clip; your smartphone can probably do that. When you apply a transition, you're asking Premiere Pro CC to play back the two clips involved in the transition plus the transition itself. Because the transition combines the two clips, it doesn't look like either of the original clips, so Premiere Pro CC can't simply read it off your storage. The transition has to be calculated. If Premiere Pro CC believes the transition might be too complex to render in real time, you'll see a yellow or red render bar over the transition.

You usually don't need to be too concerned about a yellow render bar. But if the transition isn't playing back smoothly and you would like it to, select the transition and choose Sequence > Render Selection. This creates a new preview file that Premiere Pro CC can read directly from storage and quickly display, so the render bar becomes green.

Varying Clip Speed Over Time

★ ACA Objective 4.4

 Video 2.15 Time Remapping

You've probably seen action videos and TV commercials where clips are not only sped up or slowed down but where the speed changes over time within a single clip. For example, you see a snowboarder start at a normal or faster playback speed, and then the speed changes to slow motion. You can do this in Premiere Pro CC using the **Time Remapping** feature. You can also freeze a video clip at any frame.

Varying clip speed with Time Remapping

When you worked with audio levels, you learned how keyframes on a track's rubber band let you change values over time. To work with the Time Remapping feature, you need to be able to see video keyframes and the clip rubber band:

1. In the Project panel, double-click the B-roll video you want to use. Video 2.13 uses vid-intotheshoot.mp4.
2. Set good In and Out points for the clip in the Source window to create a segment about 5 or 6 seconds long, and drag the video portion to track V2 in the timeline.
3. In the Timeline panel, expand track V2, the video track containing the clip you want to adjust.

   As with the audio track you edited earlier, you want to increase the height of the video track so that there's more room to manipulate its rubber band.
4. Click the Timeline Display Settings icon (the wrench) and make sure Show Video Keyframes is selected (**Figure 2.33**).

Figure 2.33 Choosing Show Video Keyframes with a video track expanded

5. Right-click (Windows) or Control-click (macOS) the fx badge (![fx]) of the video clip you want to adjust, and choose Time Remapping > Speed (**Figure 2.34**). Now the rubber band over the clip controls clip speed.

You can also right-click the video (not the badge) and choose Show Clip Keyframes > Time Remapping > Speed, as demonstrated in Video 2.15.

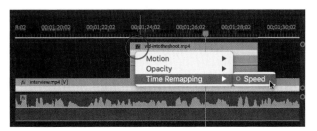

Figure 2.34 Right-click a clip's fx badge to access its Time Remapping settings.

Now you're ready to use time remapping. You'll add keyframes where you want the speed changes to occur.

To apply time remapping:

1. Do one of the following to create a keyframe:
   - With the Selection tool, Ctrl-click (Windows) or Command-click (macOS) the clip rubber band where you want to add a keyframe.
   - Move the playhead to the time where you want to start changing clip speed, and in the Effect Controls window, click the Add Keyframe button to the left of the track (**Figure 2.35**).

Figure 2.35 Clicking the Add Keyframe button in the Effect Controls window, for the clip selected in the Timeline panel

2. Repeat step 1 to create another keyframe where you want clip playback speed to change again.
3. Position the pointer over the segment of the rubber band that's between the two keyframes you added, and drag up or down to change the playback speed of just that segment (**Figure 2.36**).

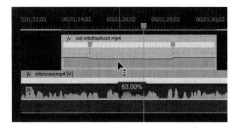

Figure 2.36 Changing the speed of the segment between keyframes

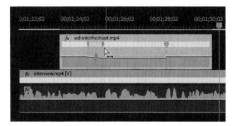

Figure 2.37 Split a keyframe to create a speed ramp.

Right now, the clip changes instantly from one speed to another. To create a smooth transition from one speed to another, you can create a speed ramp.

4 Drag either side of a single keyframe to split the keyframe and create a speed ramp. To reposition the entire split keyframe, Alt-drag (Windows) or Option-drag (macOS) the keyframe. The speed ramp extends in both directions, centered on the original keyframe time (**Figure 2.37**).

5 To adjust the acceleration or deceleration of the speed ramp, drag the blue point inside the split keyframe or its handles.

6 Play back the sequence to see the results, and make any necessary adjustments.

> **TIP**
> *You can also edit Time Remapping options and keyframes in the Effect Controls window.*

Freezing a frame while time remapping

When you want to gradually slow clip motion to a halt, you can freeze the action at a keyframe.

1 In the Project panel, double-click the B-roll video you want to use. Video 2.15 uses vid-jumpBlue.mp4 for this exercise.

2 Set good In and Out points for the clip in the Source window to create a segment about 2 or 3 seconds long, and drag the video portion to track V2 in the timeline.

3 Move the playhead to the frame showing the peak of the snowboarder's jump.

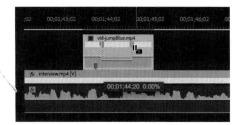

Figure 2.38 The Freeze Frame icon for Time Remapping

4 Add a Time Remapping keyframe using any of the methods you've learned.

5 Ctrl+Alt-drag (Windows) or Command+Option-drag (macOS) half of the Time Remapping keyframe, using the duration display in the tooltip to see the current duration of the freeze frame. You'll see the freeze frame icon (**Figure 2.38**) if you're doing it correctly.

6 Play back the sequence to see the results, and make any necessary adjustments.

> **MORE WAYS TO FREEZE A FRAME**
>
> The freeze-frame technique covered in this exercise is not the only way to stop motion at a specific frame. Premiere Pro CC offers other ways to freeze a frame, and you might prefer them when you aren't also time remapping a clip.
>
> **Frame Hold.** Move the playhead to a frame and then choose Clip > Video Options > Add Frame Hold. This splits the clip. The part of the clip after the split becomes a still frame matching the frame at the split point. Nothing else changes.
>
> **Frame Hold Segment.** Move the playhead to a frame and then choose Clip > Video Options > Insert Frame Hold. This splits the clip and adds a new still-frame clip segment matching the last frame before the point where you split the clip.
>
> The difference between the two options is that the Frame Hold Segment option resumes motion after the still segment. In other words, Frame Hold replaces the clip after the split point with a still frame, and Frame Hold Segment plays the entire clip but inserts a still segment in the middle, at the split point.
>
> **Export a still frame.** In the Source or Program Monitor, click the Export Frame button to save the current frame as a still image file. You can customize the image name, format, and file location in the Export Frame dialog box that appears, and there is an Import Into Project option to save you the step of bringing the exported image into the project. You can then add that still image to any sequence.

Using Markers

★ *ACA Objective 2.3*

 *Video 2.16
Markers*

As a sequence becomes more complicated, you'll probably want to mark frames for further edits at a later time. For example, you might want to identify times during the composition where you want to add titles, B-roll clips, or supplemental graphics. For the next media item you want to add, you can simply use the playhead, but what if there are numerous points in time where you want to indicate future work? Premiere Pro CC has the answer to that, in the form of **markers**. You've already used two types of marker—the In point and the Out point. Premiere Pro CC also has a more general type of marker that you can use to indicate a particular time for any reason.

When you worked with Lift and Extract earlier in this chapter, you learned that a sequence can have an In point and an Out point that are independent of the In and Out points in each clip in the sequence. It's the same with markers: each clip can have its own clip markers, and a sequence can have its own sequence markers. Apply clip markers in the Source Monitor (after opening the clip there), and apply sequence markers in the Timeline panel.

A clip marker can be useful for indicating where one clip should be aligned in time to another clip on a different track in a sequence. A sequence marker can be useful for edits or items that can be applied only in the Timeline panel. Both types of marker are useful for leaving notes to others on the production team.

Keep in mind that there's more than one way to mark items in a sequence. If you want to mark a frame (a point in time), use a marker. If you want to mark an entire clip or other media item, it may be better to change the color of the item in the timeline by changing its label (choose Edit > Label).

In the interview sequence, you can use markers to indicate in advance where you want to start a B-roll clip for a particular interview segment.

To add a marker to the interview sequence:

1 In the Timeline panel, drag the playhead to the time where you want to add a marker.

2 Do one of the following:

- Click the Add Marker button in the Timeline or Program panel (**Figure 2.39**).
- Right-click (Windows) or Control-click (macOS) the time ruler and choose Add Marker.
- Choose Markers > Add Marker.
- Press the M key.

NOTE

If you want to add a sequence marker, make sure no clips are selected. If a clip is selected, the marker may be added to the clip.

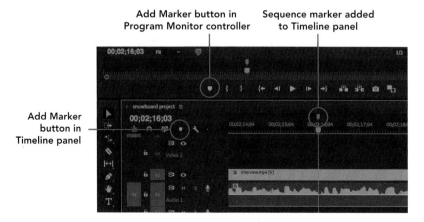

Figure 2.39 Clicking either Add Marker button will add the marker in the timeline.

3   If the marker is not at exactly the intended time, drag it along the time ruler.

You can drag to reposition sequence markers in the Timeline panel and clip markers in the Source panel.

A marker isn't just an icon. You can add information to a marker. For example, you can remind yourself or your colleagues about tasks that need to be done at that time.

To edit a marker:

1   Open the Edit Marker dialog box by doing one of the following:
    - Double-click the marker.
    - Right-click (Windows) or Control-click (macOS) the marker and choose Edit Marker.
    - If the playhead is at the marker, choose Marker > Edit Marker or press the M key.

2   Change the marker information you want to edit (**Figure 2.40**), such as the Name or Marker Color.

3   Click OK.

Pressing the M key is the easiest way to add a marker while a clip or sequence is playing.

Figure 2.40 Annotate a marker in the Edit Marker dialog box.

Adding Titles

 ACA Objective 4.2

▶ *Video 2.17*
Essential Graphics

In this exercise you'll create a **lower-third** title. As you might guess from the name, lower-third titles are conventionally set in the bottom one-third of the screen. The term has become somewhat generic since many lower-third graphics today don't occupy exactly one-third of the screen. Now the term tends to describe any title along the bottom of the screen, typically used to identify a person, or to provide context such as a location in the scene or a date and time.

This isn't the first time you've built a title, so some of what follows will be review and practice, but with new skills for you to learn. You'll start by adding a title based on a **motion graphics template**, a title designed in advance so that all you have to do is fill in your own text. In Premiere Pro CC you can choose from many predesigned motion graphics templates that you can use instead of designing them yourself.

Adding a title

To add a title from a template:

1. Switch to the Graphics workspace.

   The Graphics workspace contains the Essential Graphics panel, which you'll use to create and edit a title.

2. Move the playhead to the beginning of the sequence.

3. If the Safe Margins button hasn't been added to the Program panel controller, click the wrench icon in the Program panel and choose Safe Margins (**Figure 2.41**).

> **TIP**
>
> A shortcut for moving the playhead to the beginning of the sequence is pressing the Home key. On some keyboards without a Home key, pressing Fn+Left Arrow key functions as a Home key. The End key (or Fn+Right Arrow key) goes to the end of a sequence.

Figure 2.41 The Graphics workspace; click the wrench icon to choose the Safe Margins command.

4. In the Essential Graphics panel, make sure the Browse tab is active, and double-click the Lower Thirds folder. A number of templates appear for lower-third title graphics.

5. Drag the Classic Lower Third One Line template from the Essential Graphics panel to track V2 at the beginning of the sequence in the Timeline panel (**Figure 2.42**).

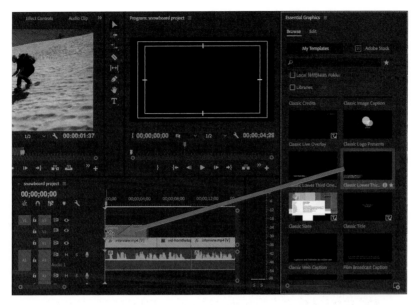

Figure 2.42 Dragging a motion graphics template to track V2 in the sequence

6. If the Resolve Fonts dialog box appears, select any fonts that need resolving, and click OK.

   If your Creative Cloud account does not have access to Adobe **Typekit**, it may not be possible to resolve fonts. For this lesson, that's okay; simply click Cancel. At any time, you can change the font to one that's installed on your system.

7. Play or scrub the beginning of the sequence to preview the template. The graphic and text of the title slide in from the left.

The title template has graphics and placeholder text for you to replace with your own. You'll do that next.

Editing a title

To edit the title text:

1. With the Selection tool (▶), click to select the title in the Timeline panel.

   The Essential Graphics panel automatically switches to the Edit tab and now displays the properties of the title. This title consists of six layers, with a text layer on top. If you select a layer in the Edit tab, that layer's properties appear below the layer list.

Chapter 2 Editing an Interview **135**

Figure 2.43 The Graphics workspace when editing titles

A **Effect Controls panel: Edit animation keyframes of selected title**
B **Program panel: Preview selected title**
C **Essential Graphics panel: Edit properties of selected title**
D **Timeline panel: Selected title**

If you display the Effect Controls panel, which is docked with the panels in the upper-left corner of the Graphics workspace, you can view the keyframes that control the animation of the layers that make up the title (**Figure 2.43**). You don't have to edit the animation, but if you wanted to, you could do so here.

2 Make sure the text layer (the top layer) is selected in the Edit tab of the Essential Graphics panel.

3 Select the Text tool, highlight the text in the Project panel, and type your replacement text. If you're following Video 2.17, type **Joe Dockery**.

To edit other title properties, make sure the title is selected in the Timeline panel, and then do any of the following:

- Edit the font, font size, and other type properties in the Edit tab of the Essential Graphics panel. If the properties are not visible, select the text layer in the layer list.
- Reposition the title by using the Selection tool to drag the title layer in the Program panel.
- Resize the title area (not the text) by using the Selection tool to drag any handle on the title layer bounding box in the Program panel.

When you want to design your own title, instead of starting from a motion graphics template you should start with the Text tool or a shape tool. You can follow the example in Video 2.17, follow the different example in the following steps, or design your own title for any part of the interview.

To create a title without a template:

1. Move the playhead to the time when you want to add the title.
2. Select the Text tool, and then click or drag in the Project window to create a new text layer.
3. Type the text for your title.

As soon as you create a new text layer, it appears selected on an empty track, and its properties appear in the Essential Graphics panel and in the Edit tab of the Effect Controls panel.

Figure 2.44 Selecting the Rectangle tool in the Tools panel

To add a shape to the title:

1. In the Tools panel, select the Pen, Rectangle (**Figure 2.44**), or Ellipse tool.

    Those three tools are grouped in the Tools panel, so only one of them is visible at a time. For example, if you can't see the Rectangle tool, that means either the Pen or Ellipse tool is the one that's visible in the Tools panel; click and hold it to reveal the Rectangle tool.

2. Create a shape by dragging the Rectangle or Ellipse tool, or clicking the Pen tool.

    The Pen tool is more challenging to use, and a complete chapter on it is outside the scope of this book. With the Pen tool, you can create corners and curves using combinations of clicking and dragging. You are encouraged to learn more about the Pen tool, because it's used in many Adobe applications.

Figure 2.45 The layer list in the Essential Graphics panel represents the visual stacking order of title elements.

3 If needed, change the stacking order of the shape relative to other title items by dragging items up or down in the layer list in the Essential Graphics panel (**Figure 2.45**).

   For example, if a shape is covering the text, go to the Edit tab in the Essential Graphics panel and drag the layers so that the text is above the shape.

4 With the shape selected, adjust its properties as needed in the Essential Graphics panel.

   For example, you can change its fill color and opacity and apply a drop shadow.

Saving a title as a motion graphics template

Many titles are used more than once. In a documentary, you might design a lower-third title to identify an interview subject and use that design for all the interview subjects in the video. You can save a title as your own motion graphics template. At any time, you can add another instance of that template to a sequence, as easily as you used the built-in Classic Lower Third template earlier.

To save a title as a motion graphics template:

1 Choose Graphics > Export As Motion Graphics Template.

   If the command is not available, make sure the Timeline panel is active and the title is selected.

2 Enter a name, select a destination and additional options (**Figure 2.46**), and click OK.

Figure 2.46 Creating a motion graphics template

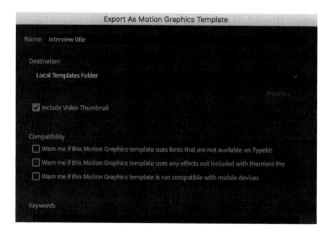

138 Learn Adobe Premiere Pro CC for Video Communication

In the Destination menu, the items below the menu divider are the Creative Cloud libraries associated with your Creative Cloud account, so they may not be available if your account doesn't have access to Creative Cloud libraries.

> **IMPORTING GRAPHICS WITH TRANSPARENT BACKGROUNDS**
>
> If you want an imported graphic to have a transparent background, the graphic must meet the following requirements:
>
> - The graphic must be created with areas that are fully transparent, not white. If you create or open the graphic in Adobe Photoshop, transparent areas appear as a checkerboard pattern behind the graphics and text layers. Note that if a graphic in Photoshop has a locked layer named Background at the bottom of the Layers panel, its background is not transparent. In Adobe Illustrator, any area not covered by an object is transparent.
> - The graphic must be saved in a file format that allows transparency.
>
> Some of the most popular and highest-quality graphics file formats that support transparency are Adobe Illustrator (with the extension .ai), Adobe Photoshop (.psd), Tagged Image File Format (.tif or .tiff), and **Portable Network Graphics (.png)**. AI format is a vector format, so it scales smoothly to any size.
>
> **Graphics Interchange Format (.gif)** is capable of only one level of transparency; the other formats can store up to 256 levels of transparency for smooth, anti-aliased edges. For this reason, it's better to use the other formats.
>
> The JPEG (Joint Photographic Experts Group) file format (using the .jpg filename extension) cannot store transparency; it always has a solid background.

Stabilizing a Shaky Clip

ACA Objective 4.5

ACA Objective 4.6

▶ *Video 2.18*
Stabilize Your Video

Steady footage with smooth camera moves gives a professional look to a production, but in-camera stabilization can do only so much, and high-quality camera stabilization hardware isn't always practical to have on hand when recording. When your clips could benefit from stabilizing in postproduction, you can apply the Warp Stabilizer effect in Premiere Pro CC.

Applying Warp Stabilizer

Try applying Warp Stabilizer to a B-roll clip that looks shaky, such as vid-fromthetop.mp4. Applying Warp Stabilizer is like applying any other effect. Find it in the Effects panel (it's in the Video Effects > Distort group) and drag it onto a clip in a sequence in the timeline.

However, Premiere Pro CC won't let you apply Warp Stabilizer to a clip that has an effect such as Speed/Duration applied to it. In Video 2.18, the Speed effect is removed so that Warp Stabilizer can be used.

Warp Stabilizer is also unavailable for clips used in sequences with settings that don't match the clips' resolution.

Stabilization is a two-stage process. First, Warp Stabilizer analyzes the footage to determine the degree and direction of instability. Then it applies the precise amount of stabilization necessary to cancel out the instability that was detected. Both the analysis and stabilization stages are processor-intensive and can take a lot of time. Exactly how much depends on the length and resolution of the clip you're stabilizing.

When Warp Stabilizer is in the middle of either of its two stages, it displays a banner across the Program Monitor when the playhead is displaying a clip that Warp Stabilizer is processing. The banner says either "Analyzing" or "Stabilizing." You can also monitor Warp Stabilizer progress in the Effect Controls panel when the clip is selected. Warp Stabilizer processes in the background, so you can keep working on other parts of your project.

> **TIP**
> It is possible to combine effects such as Speed and Warp Stabilizer using an advanced technique: create a sequence from the clip alone, stabilize the clip there, and then add that sequence to your main sequence.

> **TIP**
> When you apply processor-intensive effects such as Warp Stabilizer to a clip, that clip may take longer to preview. For faster previewing, you may want to disable processor-intensive effects while you complete other edits, or apply those effects after completing most other edits.

Customizing Warp Stabilizer

Although you can simply apply Warp Stabilizer and use the result, there are a few key settings you might want to pay attention to in the Effect Controls panel (**Figure 2.47**):

- **Result.** You should generally leave this option set to Smooth Motion, but if you want to make the shot look like the camera never moved, select No Motion.
- **Method.** If the default Subspace Warp method causes unwanted effects, you can try choosing a simpler method. Position is the simplest.
- **Framing.** Warp Stabilizer can use different methods to handle edges when stabilizing. If the default Stabilize, Crop, Auto-Scale framing causes too much

variation in scaling, you can try choosing a simpler framing. Stabilize Only is the simplest. Stabilize, Synthesize Edges will create fill areas for empty areas resulting from shifting and scaling the frame during stabilization; how well this synthesis works depends on the content in the frames.

Figure 2.47 Warp Stabilizer settings in the Effect Controls panel (left), as Warp Stabilizer analyzes the clip in the Project panel (right)

In addition to these settings, you can explore the Advanced settings to use options that tune the balance between smoothness and the compromises inherent in scaling and edge synthesis.

Camera shake is easier to remove when there's a small amount of it. The greater the shake, the more the frame has to be scaled or shifted to compensate, and the more image quality will be affected. Warp Stabilizer may not be able to produce satisfactory results with extreme camera shake, such as holding an unstabilized camera with one hand while walking.

> **TIP**
>
> *You can copy or paste effects between clips. Choose Edit > Copy for the first selected clip; then select another clip and choose Edit > Paste Attributes (not Paste). If you paste Warp Stabilizer, remember to have it reanalyze the new clip.*

★ ACA Objective 4.4
★ ACA Objective 4.7

▶ Video 2.19
Merging Clips

Merging Separate Video and Audio Files

When you want to create a professional video production, you probably won't be satisfied with the sound quality of the microphone built into your video camera. You'll do what is standard practice in professional video production: record video and audio separately. This lets you use a dedicated high-quality audio recorder, and you can choose the best microphone for the job. You also get the freedom of placing the microphone in position for the best-quality audio, which is often quite different than the best position for the camera. For example, you can place a lapel mic on a speaking subject, which will record much clearer dialogue with much less background noise than an on-camera microphone.

Of course, recording video and audio separately means you have to put them together later, and they have to be perfectly synchronized. Fortunately, Premiere Pro CC provides a quick and easy way to do this.

To merge separate video and audio files:

1. In Premiere Pro CC, use any method you've learned to import the Merge Clips folder from the project2_snowboarding folder on your desktop.

2. In the Project panel, open the Merge Clips bin and select both of the clips.

3. Choose Clip > Merge Clips, or right-click (Windows) or Control-click (macOS) and choose Merge Clips.

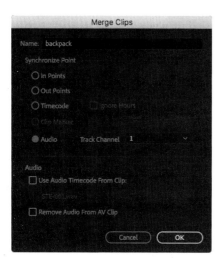

Figure 2.48 The Merge Clips dialog box

4. In the Merge Clips dialog box (**Figure 2.48**), do the following:

  - For Name, enter **backpack**.

  - For Synchronize Point, select Audio. This option is automatic; Premiere Pro CC analyzes the audio in both clips and lines them up so that they match. All of the other options require manual preparation; for example, you can synchronize by timecode only if you're sure that you synchronized the audio recorder and video camera timecode settings before recording.

  - For Audio, select Remove Audio From AV Clip. This deletes the audio that the video camera recorded. You normally want to select this, because the intention is to replace the unwanted video camera sound with the high-quality sound from the audio recorder.

142   Learn Adobe Premiere Pro CC for Video Communication

5 Click OK.

The video clip with audio from the camera and the sound clip from the audio recorder are merged into a single clip, and the audio recorder sound replaces the video clip's sound. Now you can use the merged clip that contains the high-quality audio for the video.

Exporting with Adobe Media Encoder CC

★ ACA Objective 5.1

★ ACA Objective 5.2

▶ Video 2.20
Exporting with Adobe Media Encoder

When your sequences are complete and ready to be rendered out to final video files, you can export them. When you export, not only does Premiere Pro CC have to assemble all the components of a sequence into a single document, it might also have to convert all the media in the sequence to a different format and compress data to keep the file size down. Doing those tasks for every single frame will keep your computer very busy and can take a long time. Higher resolutions (such as 4K frames), heavy use of effects, and longer sequences can extend the exporting time further. That time can be cut down by using a computer with more RAM, faster drives, or a more powerful graphics card that's compatible with the Mercury Graphics Engine.

When a sequence takes many minutes or even hours to export, you don't want to be unable to use Premiere Pro CC for that entire time. For this reason, you can send an exported sequence to Adobe Media Encoder CC. It can process a queue of sequences in the background so that you can continue working on other projects in Premiere Pro CC.

For this exercise, you'll export a sequence to Adobe Media Encoder CC with settings that are appropriate for YouTube. You've already done these steps in Chapter 1; the difference here is that the sequence will be sent to Adobe Media Encoder CC for background rendering instead of being rendered directly in Premiere Pro CC.

To export a sequence for YouTube:

1 Make sure the sequence you want to export is either active in the timeline or selected in the Project panel or bin.

2 Play the sequence at least once to see whether anything needs to be fixed before exporting.

3 If you want to export just part of the sequence, set sequence In and Out points.

TIP

The default output name is the sequence name, so you won't have to correct the output name every time if you give your sequences the same names you intend for the final exported video files.

Figure 2.49 The Queue button sends the sequence to Adobe Media Encoder CC.

4. Choose File > Export > Media.
5. Click the Format pull-down menu and choose H.264.
6. Click the Preset pull-down menu and choose YouTube 720p HD.
7. Click the blue Output Name text to set the location and filename for the exported video.

   Make sure it will be saved to the Exports folder that you set up for this project.
8. Click the Queue button (**Figure 2.49**). This will send the sequence to the render queue in Adobe Media Encoder CC.

9. If you have another sequence that you'd like to queue for rendering, such as another version of the sequence, do so by repeating steps 1 through 7 in Premiere Pro CC.
10. Adobe Media Encoder CC should start (**Figure 2.50**). If it starts in the background, switch to it.

    In the Queue panel you can see all the sequences or other media that you've sent to Media Encoder CC. If you want, you can change the processing order of the items by dragging them up or down in the list.

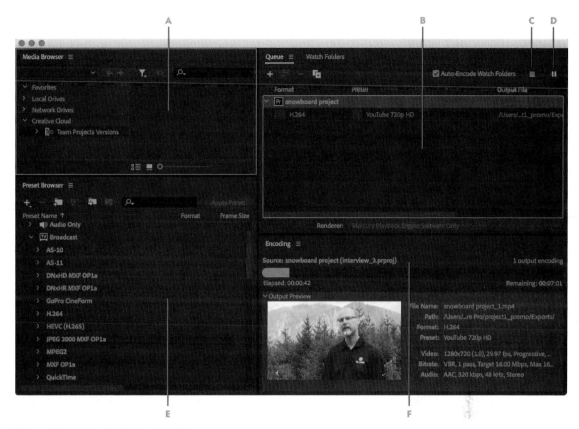

Figure 2.50 Adobe Media Encoder CC

A **Media Browser panel:** As in Premiere Pro CC, use it to import media

B **Queue panel:** Lists sequences and clips being processed or waiting

C **Stop Queue button**

D **Pause Queue button**

E **Preset Browser panel:** Browse presets to be applied to items in the queue

F **Encoding panel:** Preview and monitor progress of jobs currently processing

You can use Media Encoder CC as a stand-alone compressor and **transcoder** (format converter). If you have a folder full of clips that need to be converted, you can use the Media Browser to add them to the Media Encoder queue and assign a preset to them from the Preset Browser.

11 Click the green Start Queue button. Media Encoder CC begins processing the items in the queue. The Start Queue button becomes a Pause Queue button while the queue is being processed.

Chapter 2 Editing an Interview **145**

TIP

If you need to change the export settings, click the Preset arrow or text for the item in Media Encoder CC. To change the filename or export location, click the Output File name. You can change settings at any time before the item starts rendering.

Because Media Encoder CC processes in the background, you can switch to other programs while it's working. However, the Export queue will process faster if you run as few other programs as possible while it's processing, to avoid having programs compete for the processing power of your computer. Also, other programs may run slower while Media Encoder CC is processing.

If you realize there is a more important task to complete, you can click the Pause Queue button.

12 When the last export in the queue is complete, you can check your exports and exit Media Encoder CC.

TIP

If you set sequence In and Out points because you wanted to export a range of frames instead of the whole sequence, in the Export Settings dialog box you can select Sequence In/Out for Source Range.

Challenge: Mini-Documentary

Create a short documentary about an interesting person in your life. Maybe someone in your family has had a fascinating career path, traveled the world, or served with honor. Or do you know a great teacher or a friend with an unusual hobby?

As you plan your documentary, remember Joe Dockery's Keys to Success (from the video):

▶ *Video 2.21*
Mini-documentary Challenge

- Keep it short.
- Plan. When planning interview questions, keep the questions open-ended. Ask interview subjects to describe and explain their experiences and actions.
- Record good-quality video and audio. For video, light subjects in a flattering way. For audio, keep the microphone close to the person speaking and monitor the audio using headphones. Good monitoring helps you catch problems while you can fix them, since you don't want to have to do an interview twice.
- For B-roll, collect and scan photographs, medals, and other media and mementos.
- Share your work with the world...build that portfolio!

CHAPTER OBJECTIVES

Chapter Learning Objectives

- Finding media on stock websites.
- Learning more details about the Preferences dialog box.
- Managing links to imported media.
- Using the Properties command.
- Editing vertical video.
- Editing a multicam sequence.
- Sweetening different types of audio.
- Using an adjustment layer.
- Adding rolling credits.
- Recording a voiceover.
- Using proxies.
- Exporting multiple sequences.
- Cleaning up and archiving a completed project.

Chapter ACA Objectives

For full descriptions of the objectives, see the table on pages 279–283.

DOMAIN 1.0
SETTING PROJECT REQUIREMENTS
1.1, 1.2, 1.3, 1.4

DOMAIN 2.0
UNDERSTANDING DIGITAL VIDEO
2.1, 2.2, 2.3, 2.4

DOMAIN 3.0
ORGANIZATION OF VIDEO PROJECTS
3.1

DOMAIN 4.0
CREATE AND MODIFY VISUAL ELEMENTS
4.1, 4.2, 4.3, 4.5, 4.7

DOMAIN 5.0
PUBLISHING DIGITAL MEDIA
5.1, 5.2

CHAPTER 3

Editing an Action Scene

The third Adobe Premiere Pro CC project is a public service announcement (PSA) for a school, designed as a fast-moving action scene to get students' attention.

Getting Ready in Preproduction

A school needs a video that encourages students to stay in shape and to get to class on time, so that's what you'll create in this chapter.

As you've learned, production starts only after the project requirements are clearly understood, so review those before you begin:

- **Client:** School district
- **Target audience:** High school students
- **Purpose:** Encourage students to stay in shape and to be in class on time.
- **Deliverable:** The client expects a fast-paced 15- to 30-second video of action scenes. To load quickly online, the video should be in H.264 Vimeo 720p HD format.

★ *ACA Objective 1.1*

★ *ACA Objective 1.2*

▶ *Video 3.1 Job Requirements*

Unzipping the project files

On your desktop, unzip the project files in the project3_action.zip file so that you can work with the folders it contains. There's a MediaFiles folder and a Pre-Production folder. Here are the folders inside the MediaFiles folder:

- **Audio Clips:** This folder contains a sound effect of an electronic school bell.
- **Graphics:** The project uses one still image, which is in this folder.
- **Presets:** This folder contains an Adobe Media Encoder preset, which is a settings file that controls how a Premiere Pro sequence exports. Adobe Media Encoder preset files can be provided to project colleagues to ensure that sequences are consistently exported to the specifications required for the project. In this case, the settings in the preset are optimized for posting video on social media.
- **Project:** This is an empty folder where you'll store the Premiere Pro project file that you'll soon create.
- **Video Clips:** Here you'll find video clips from multiple cameras and a vertical video from a smartphone.

Locating your own appropriate background music for this sequence is part of the exercise. Look through online sources of free music to find a clip that supports the action. If you have the ability and the software, you can also compose your own music.

You can also look through your own digital music collection, but keep in mind that in a real-world project, you'll need to clear any music used in a video for use with the proper usage rights and possibly fees. If permission is not required, **attribution** (crediting the creator) may still be required. In many countries, all of that applies to media used in nonprofit or personal projects as well as commercial ones.

Reviewing the storyboard and shot list

▶ *Video 3.2*
Pre-Production

The project files contain a Pre-Production folder. There are a couple of items that you haven't seen before in this course.

Let's take a look at the pre-production files provided for this project:

- **shot list-action.pdf:** A shot list describes all the shots you plan to take, organized by location and setup. A shot list is valuable because it helps ensure that all shots that use a specific location and setup are anticipated and taken together. It can take a while to set up cameras and lighting for a scene, so even if there's only one location, you can use the shot list to avoid having to redo a setup for a missed shot. You really want to avoid having to come back to a location to get a missed shot, because that requires finding a time when the location, crew, and actors are all available, as well as duplicating the lighting and blocking of the original shoot. That may create unanticipated burdens on the production budget and schedule. The shot list helps maximize the efficiency of the production.

- **storyboard-action.pdf:** A storyboard is a sequence of sketches that indicates how shots should look. It's a great visual planning tool that the director, actors, and director of photography can use to agree on how to set up each shot.

You can create shot lists and storyboards using nothing more than a pencil and a sheet of paper. But you can also find websites and apps that are designed to help plan video projects. Because video projects are typically collaborative, you might consider using cloud-based tools that make it easy for multiple people to view and edit. For example, you might create a shot list using Google Sheets spreadsheet software. You might make a storyboard by pasting clip art into Google Docs or by drawing on your phone or tablet in the Adobe Photoshop Sketch mobile app.

Acquiring and Creating Media

★ ACA Objective 1.3

 Video 3.3
 Stock Media

Although a large part of video production is recording your own audio and video, it isn't always practical to create every last bit of media that a production might require. Sometimes it takes less time, money, or effort to acquire certain types of media from other sources. Or maybe your production needs media that your team does not have the ability to create.

This project needs background music for a fast action scene. There isn't any that's immediately available, but the scene doesn't require an original score. It's worth looking around for a music track that's already available to use from a stock website.

Finding media on stock websites

Many websites that offer stock photography and stock video clips also offer stock music. You can find websites offering **royalty-free** media that you can download without having to pay, but for better selection and quality you may want to explore paid options as well.

You'll see some stock media sites mentioned or shown in the videos and in this book. Keep in mind that they are only suggestions and not recommendations. Always do your own research on stock media sites to make sure they fit your specific needs. As fast-moving as the online world is, more options may have become available to you after this book was published.

The following are some free stock media sites:

- **Incompetech (incompetech.com):** This is a popular site for royalty-free music, which is one reason it's used as an example in Video 3.3. You can obtain media on this site under a selection of paid and free licenses depending on the usage, so be sure to read and abide by the license terms.
- **Unsplash (unsplash.com):** Focusing on photographs, Unsplash doesn't charge a fee, and it doesn't require attribution (although it encourages it).
- **Pixabay (pixabay.com):** Another website oriented around still images, Pixabay offers images under a Creative Commons Public Domain license (see the next section, "Types of Licenses").

Remember, even a free stock media download may require attribution. Before you download free media, always review the licensing terms and make sure you agree and adhere to them; fulfilling your end of the deal respects contributors and helps sustain the system of free media.

Here are some fee-based stock media sites:

- **Adobe Stock (stock.adobe.com):** Some Adobe Creative Cloud plans may include a certain number of Adobe Stock downloads for no additional cost. If you have a more basic Adobe Creative Cloud plan, Adobe Stock is probably not included, so downloads may incur a fee. Compared to other stock websites, a major advantage of Adobe Stock is that it integrates more deeply with Adobe Premiere Pro and other Adobe applications. Also, Adobe Stock offers more than just video, sound, and image files; for example, it offers templates for Premiere Pro titles and motion graphics.
- **Storyblocks (storyblocks.com):** Offering videos, images, and audio through a paid subscription service, Storyblocks may be useful if you're constantly busy with video projects that need additional media.
- **Triple Scoop Music (triplescoopmusic.com.com):** Focusing on music, Triple Scoop Music offers a range of fees for different uses, and it provides a viable option if you don't want to pay a regular subscription fee.

If you're working on a project for an organization, check to see if they already have a membership to a stock library.

> **TIP**
> *Always double-check the licenses, offerings, plans, and features of any stock media website before you use it, because they may have changed since this book was written.*

> **NOTE**
> *For more on media licensing and other licensing issues, see "Types of licenses" on page 264 in Chapter 6.*

Making your own media

Creating your own music doesn't mean you have to write a score and play instruments. Desktop and mobile software applications are available that make it easy to create a song by providing music loops or other music building blocks that you connect and layer however you like. Some examples are Apple GarageBand, FL Studio, and Ableton Live. Some tools even let you build a song based on mood and tempo, requiring no musical knowledge.

When you make your own media, you can share it to one of the free media websites mentioned earlier. But be as careful with terms and licenses as you would be if you were using downloaded media, and be very aware of the terms that you agree to. For example, if you agree to have your song or video distributed for free under a Creative Commons or public domain license, you may find it difficult to claim ownership or make money from that work.

Obtaining model releases

The purpose of a **model release** is to legally affirm that you have permission to depict the people in your video. It's called a release because it releases you from legal liability. The example in this chapter is a school production that involves people under 18 years of age. Since they are minors, it is necessary to obtain signed model releases from their parents.

There are numerous model release templates that you can download on the Internet, and there are also mobile apps that claim to be able to generate a legal model release.

Consider legal help

If you build a growing video production business, you can help protect your company by knowing an attorney who is familiar with media-related legal issues. A media attorney can help you understand licenses for media items you want to use, as well as licenses for your own media that you want to distribute or sell. The attorney can also review your model release to make sure it's legally valid according to the laws of your specific country and state.

★ ACA Objective 1.4

★ ACA Objective 2.2

▶ **Video 3.4**
Preferences Detailed

Taking Another Look at Preferences

You've worked with some Premiere Pro preferences earlier in this book, and it's a good time to dive a little deeper into them.

Remember that how you open the Preferences dialog box depends on whether you're using Premiere Pro CC on Windows or on a Mac (see "Opening Preferences" in Chapter 1).

The Preferences dialog box contains a large number of options, and you certainly don't need to adjust them all—only the ones that might make editing easier for you. For the same reason, they aren't all covered here. What follows is based on the tour of Preferences by Joe Dockery in Video 3.4.

If you'd like to read a more detailed description of the options in the Preferences dialog box, see https://helpx.adobe.com/premiere-pro/using/preferences.html.

Controlling general behavior

In many applications, preference settings control frequently encountered behavior that might work well for someone else but prove at odds with your preferred workflow or otherwise slow you down. By changing these behaviors to a way that you prefer, you'll be more efficient and maybe enjoy editing more. These options are in the General pane of the Preferences dialog box:

- **At Startup:** The first option in the General pane, this option controls what Premiere Pro shows you when you start the application. By default it's set to Show Start Screen, where you can create new projects or open tutorials—a good setting for a beginner. But if what you most commonly do is continue work on the project you were working on earlier, you may want to change At Startup to Open Most Recent.

- **Bins:** When you double-click a bin, do you want it to open in the same panel as its parent, or in a new tab or window? You can set that here. You can use the other Bins options to set what happens when you double-click a bin while holding down a modifier key. For example, you can set the preference so that double-clicking opens a bin in the same panel, and Alt/Option-double-clicking opens it in a new window.

Changing the appearance of Premiere Pro

The Appearance pane of the Preferences dialog box affects the presentation of the Premiere Pro user interface, including the following:

- **Brightness:** This controls whether the background behind Premiere Pro panels, windows, and dialog boxes is dark, light, or somewhere in between. Traditionally, video editing applications tend to have a dark background, but if you typically use other software with lighter backgrounds, you can adjust Premiere Pro to be more consistent with them.
- **Highlight color:** Premiere Pro highlights items in a blue color that you can't change. But you can use the Interactive Controls and Focus Indicators options to change how light and how saturated that color is for different ways that controls are presented. For example, if you like to press Tab as a shortcut to move among controls in a dialog box, the Focus Indicator color is what marks the item you've focused on with the Tab key.

Configuring audio hardware

If you're editing on a computer that isn't connected to anything, like a laptop you're using in a library or coffee shop, you probably don't need to change anything in the Audio Hardware preferences pane. But if you connect audio hardware such as a microphone or a USB audio interface, you may need to adjust Audio Hardware settings.

A common example is if you try to record through connected audio hardware and Premiere Pro doesn't seem to be receiving audio from it. You'll want to open the Audio Hardware pane and try selecting the name of your audio hardware from the Default Input menu.

Similarly, if you can't hear audio from Premiere Pro CC, one of the things you should check is whether the speakers, headphones, audio interface, or other audio output hardware is selected in the Default Output menu.

Controlling how imported media scales to the frame

When you add a clip or a still image to the Timeline and its frame size doesn't match the timeline settings for the current sequence, should the media frame size be adjusted to match the sequence's frame size? That's a question you can answer in the Media pane of the Preferences dialog box.

> **NOTE**
>
> If a clip's frame size does not match the Sequence Settings, you'll find that some effects, such as Warp Stabilizer, cannot be applied to that clip. To stabilize the clip, you'll need to stabilize it in another sequence and import it using a nested sequence, as described later in this chapter.

Specifically, the Default Media Scaling option affects how the frame size of new Timeline media is adjusted for the frame size of the Timeline:

- **None:** The frame size of media added to the Timeline won't change.
- **Scale To Frame Size:** This resamples the clip to the frame size, which is typically not recommended because it alters the clip's pixels and may result in lower image quality at some sizes.
- **Set To Frame Size:** This applies the size change using the Scale option in the Effect Controls panel, which preserves the clip's pixels and original quality. Set To Frame Size is especially preferable if you plan to enlarge the clip in the sequence as a zoom-like effect.

Adjusting Timeline options

The Timeline panel contains more of the settings that affect editing tasks that may be repetitive. By changing a setting so that it works the way you want from the beginning, you can spend less time adjusting clips and Timeline controls. Here are a few examples:

- **Video Transition Default Duration and Audio Transition Default Duration:** Suppose you've add 20 transitions to the Timeline, they all seem too long, and you know you'll need to add a lot more transitions before you finish. Instead of having to fix large numbers of transitions later, you can change Video Transition Default Duration to a lower number before you add more transitions. You can do the same for audio transitions.
- **Still Image Default Duration:** In the same way that you can control the default duration of a transition, you can also set how long still images appear in the Timeline when you add them.

You may also want to explore other options in the Timeline panel, which affect things such as whether the Timeline scrolls itself during playback, how your mouse scrolls the Timeline, and whether you want to see the Show Clip Mismatch dialog box.

Adjusting the media cache

The Media Cache Preferences panel (**Figure 3.1**) contains settings for managing rendered media that's used to save time when playing back a sequence.

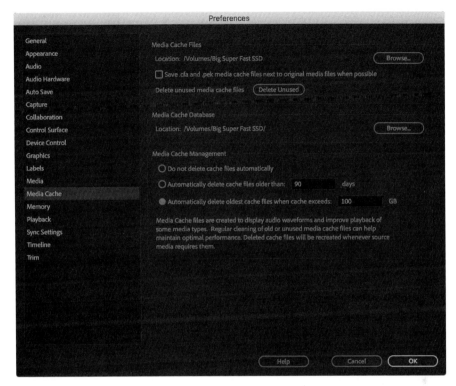

Figure 3.1 The Media Cache Preferences panel

It can take a lot of processing power and time to render your clips and edits along with all the adjustments and effects you might apply, so Premiere Pro retains frames it has already processed along with audio waveforms it's already drawn. It keeps all of that in a folder on your drive called a **media cache**. If some frames have already been rendered and no changes have been made to them, the frames can be played directly from the cache, which is usually faster than rendering them again.

For the media cache to contribute the most to editing performance, it should reside on as fast a drive as possible, such as a solid state drive (SSD). If you're editing on a computer that has only one drive, you don't have a choice; you have to use the internal drive for the media cache, which is how the Media Cache panel is set up by default. If you're editing on a workstation connected to multiple drives and one or more of them is as fast or faster than your system drive and with plenty of space available, you can use the Media Cache Files setting to put the media cache on that fast drive, freeing up space on your system drive.

On any computer, you can use the Media Cache settings to manage the space used by cache files. Eventually, the media cache may still hold cache files for a project you finished a month ago. Video files are large, so you might want to delete the cache files you don't need anymore. You can do this manually and automatically:

- To manually delete unused cache files, click Delete Unused.
- To let Premiere Pro delete cache files by age, select "Automatically delete cache files older than __ days," and enter the number of days you want cache files to be kept before deletion.
- To let Premiere Pro delete cache files before they take up too much space, select "Automatically delete oldest cache files when cache exceeds __ GB," and enter the number of gigabytes to limit the size of the media cache.

If you're using a desktop computer with large drives, you may choose to delete cache files manually after keeping an eye on available storage space. But if you're using a laptop computer or older computer with a small internal drive, you may prefer to use the automatic options so that Premiere Pro continually limits the size of the media cache before it uses too much space.

★ ACA Objective 2.1
★ ACA Objective 2.4

Setting Up the Action Scene Project

Earlier you unzipped the project3_action.zip file and took a look at the included files. Now you'll set up the project itself.

Create a new project

Now you'll set up the interview project.

1. Start Premiere Pro, and when the Start screen appears, click New Project.
2. Enter a name. In the videos, the name of the project is **Tardy PSA** because the purpose of the video is to encourage students not to be late for class, but you can use a different name if you want.
3. In the New Project dialog box, click Browse, navigate to the project3_action folder, then to the Project subfolder, and save this project there.

   The folder you applied becomes the default folder for all of the locations in the Scratch Disks tab, and you can leave them as is for this project. You also don't need to change any settings for the Ingest Settings tab for this project. In the Ingest Settings tab, Ingest should be disabled.

4   In the General tab, select Display The Project Item Name And Label Color For All Instances.

5   You can leave the General and Scratch Disks tabs at their default settings unless you have a special location for scratch disks, as discussed in Chapter 1.

6   Click OK to finish creating the interview project.

7   Save the project.

Importing Files and Maintaining Links ★ ACA Objective 2.4

▶ **Video 3.5**
Importing and Linking Media

In Chapter 1, you read about how a Premiere Pro project does not copy imported media files into the project file; instead, it remembers the folder path to the media files (it links to them) so that it can read and play those media files for editing and rendering. Because media files are stored externally, they must be available at all times when the project is being edited. If a media item isn't present at its last known storage location when Premiere Pro looks for them, the link is broken and that media item can't be displayed.

A link to a media file can be broken when the file is moved or renamed. A link can also be broken when a media file is imported from a location that disappears, for example, when a video file is imported from a camera card that is later removed from the computer, or imported from an external storage drive that is later disconnected and assigned a different drive letter when reconnected.

To avoid these problems, import media into Premiere Pro from folders on drives that will always be connected when you edit, and do your best to not move or rename files that have already been imported. Make sure you've committed to filenames and locations before importing.

Importing from a reliable location

The project examples you have seen always show media files being copied or moved into a folder created for the project. This is because you want the media files to be in a location that will be immediately accessible to Premiere Pro during the entire time you'll be editing.

Don't import media directly from a Downloads folder, a camera card, a USB stick drive, or other removable media. Copy or move media from those locations to your project folder before importing so that the media stays with the project after the camera card or other removable media is disconnected from the computer. It's okay to import media from an external storage drive *if* that drive will always be connected to the computer while you're working on the video project.

Keeping all project media together is especially important when you work with others. For example, if you import a free music file from your Downloads folder and then you hand the project files to your colleague, he will be missing the music file if it's still stuck in the Downloads folder back on your computer. If you put the music file into the same folder as the rest of the project media, it will be there when you hand the project folder to your colleague.

For this project, all the media is already contained within the project3_action folder, in the MediaFiles folder, so the media is in the correct place for importing. Specifically, you'll import the contents of the 4K Clip, Audio Clips, Graphics, and Video Clips folders in the MediaFiles folder. You won't import the Presets and Project folders.

1 Import the contents of the 4K Clip, Audio Clips, Graphics, and Video Clips folders into the Project panel, using any of the techniques you've learned:

   Use the File > Import command.

   Use the Media Browser.

   Drag and drop the folders from the desktop to the Project panel.

2 In the Project panel, create a Sequences bin because you'll be using more than one sequence.

3 Save the project.

Figure 3.2 The Media Offline screen

Relinking offline media

When Premiere Pro tries to play a media file that can't be located, it displays a red screen that says Media Offline (**Figure 3.2**). That means the link to the file is broken—the media file doesn't exist at the folder path and name that it had when it was imported.

Before you do anything, stop and think about whether the media is on an external storage drive that simply needs to be connected to the computer. This may be the case if you regularly use an external drive as a media drive. When you connect an external drive containing media files used in a project, the media that was marked offline should be restored to online status as soon as Premiere Pro can see the media files at the same folder paths that were recorded in the project.

If a media file is offline because it was moved or renamed, manually relink to it:

1 Choose File > Link Media (**Figure 3.3**)

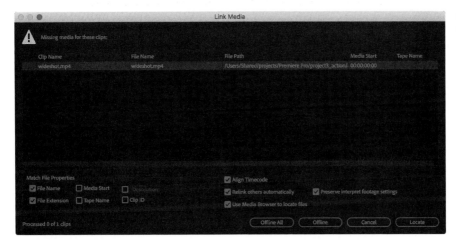

Figure 3.3 Link Media dialog box

You can choose the Link Media command by right-clicking an offline media item.

The Link Media dialog box opens, listing all offline media. It reminds you what filename you're looking for and its last recorded folder path.

2 Select a media item that you want to relink and click Locate.

If multiple media items are missing, select the Relink Others Automatically check box to cut down on the amount of searching you have to do.

If you click Offline All or Offline, the paths to the selected offline media will not be resolved at this time, and Premiere Pro will continue to display the Media Offline screen for any media still offline.

3 In the Locate dialog box (**Figure 3.4**), find and select the missing file, and click OK.

Figure 3.4 The Locate File dialog box

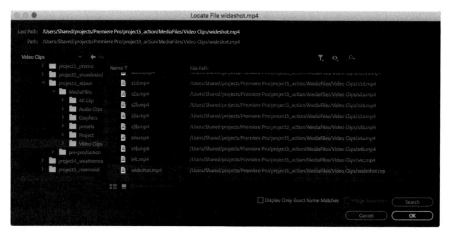

The Locate File dialog box works a lot like the Media Browser, offering tools that you can use to try to find the missing item. You can navigate the folder tree on the left and use the filter and search features above the right corner of the thumbnail grid.

If you're sure the item wasn't renamed, you can select the option Display Only Exact Name Matches to help find the missing item more quickly out of a long list of files.

If a project's media is moved, renamed, or otherwise becomes unavailable when that project is closed, Premiere Pro will alert you to the missing media the next time you open the project. The Link Media dialog box will be available for you to relink any missing media before continuing to open the project.

Be careful when Ingest is enabled

The Ingest option in the Media Browser (**Figure 3.5**) can save time, because it can copy and preprocess media as it imports the media, in ways that haven't been discussed yet such as creating proxies and transcoding (converting formats). But the Ingest feature has options that can change the folder locations of media, and Ingest stays enabled until you turn it off. This combination can sometimes lead to unexpected results. If you leave Ingest enabled, make sure you are aware of how it's set up. Otherwise, media files you're ingesting for a current project might be using Ingest settings and drive locations that were set up for an earlier project, creating copies of media files in locations you weren't expecting, possibly outside of the folder you've set up for the current project.

Figure 3.5 The Ingest option in the Media Browser

Inspecting the Properties of a Clip

★ ACA Objective 1.4
★ ACA Objective 2.4

As you work with a media item, you might need to know more about it. For example, if it's a video file, you might want to find out the format and frame rate of a clip. The Properties command may have the answers you need.

To use the Properties command, select a media item and choose File > Properties. The Properties command is also available if you right-click (Windows) or Control-click (macOS) a clip.

▶ **Video 3.6**
Properties

The Properties panel (**Figure 3.6**) opens, displaying a list of properties. The items in the list can change depending on the type and format of the selected media item. Video clips, still images, and audio clips display different properties, but you also might see a different list of properties for clips captured in different formats.

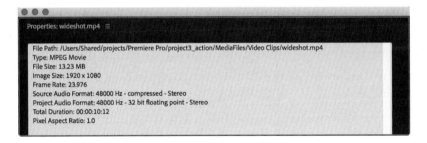

Figure 3.6 The Properties panel

You might see the following list of properties for an HD clip captured with a smartphone:

- **File Path:** The file path describes the hierarchy of folders leading to the file. This path is how Premiere Pro remembers the location of a media file. If the file is on a different drive, the drive name or letter is included at the beginning of the path.
- **Type:** The type is the file type of the media item.
- **File Size:** The file size is how much space the media item occupies on your storage drive.
- **Image Size:** The image size is the size of the video or still frame in pixels. For example, 1920x1080 pixels is the image size of a 1080p (2K) high-definition video frame.
- **Frame Rate:** The frame rate is the rate at which the frames of the media item are played back, in frames per second—for example, 29.97 frames per second.

TIP
You can change how Premiere Pro interprets some clip properties, such as frame rate and pixel aspect ratio, by using the Clip > Modify > Interpret Footage command.

- **Audio Format:** The audio format describes the specifications of the audio. For example, "44100 Hz - compressed - Mono" means it has a sampling rate of 44,100 hertz, is compressed audio, and is monophonic (has one audio channel).
- **Total Duration:** The total duration is the duration of the entire clip, when In and Out points are not set.
- **Pixel Aspect Ratio:** You may already know that a video frame has an aspect ratio—the ratio of its width to its height. Though many image and video formats use square pixels (1.0), some formats use pixels that are rectangular in shape. If a pixel has a nonsquare aspect ratio, Premiere Pro can be set to compensate so that the image is not distorted.
- **Variable Frame Rate Detected:** Though professional cameras record video at a precise and constant frame rate, some consumer devices record video at a frame rate that can vary. This can include video from smartphones and game-play recording devices, as well Skype and some other streaming platforms that can also record an archive of your streamed video. If variable frame rate is not handled properly, video and audio can drift out of sync. When Premiere Pro detects that a video was recorded with a variable frame rate, it can automatically attempt to keep video and audio in sync.

Remember that you can mix clips of different types in a sequence. It is not necessary for all clips to have exactly the same properties.

★ ACA Objective 1.4
★ ACA Objective 4.1
★ ACA Objective 4.3
★ ACA Objective 4.7

▶ Video 3.7
Rough Cut

Starting a Rough Cut

The project you worked with in Chapter 2 was straightforward to edit, because it was an interview. All you had to do was string together the clips of the person being interviewed. This project is different because it's based on a conceptual storyline. The video clips were captured according to the storyboard and shot list you examined earlier, and now you can use the storyboard as a guide for building the rough cut. After you add captured clips to the sequence in the order indicated by the storyboard, you'll have a good foundation for the edits, audio, and effects you'll apply later.

Using the storyboard as a guide

The storyboard scenes can guide file naming as well as the order of clips in the sequence.

Open storyboard-action.pdf and look at what's written above the top-left corner of each frame (**Figure 3.7**). For example, the first one says s1A, and the second one says s1B. The storyboard was set up this way:

- In this instance, "s" means scene or sequence (a conceptual sequence of shots, not a Premiere Pro sequence).
- The number is the shot in the sequence.
- The letter indicates which camera should be used to capture that clip.

Therefore, s1A means scene 1, shot 1, captured by camera A.

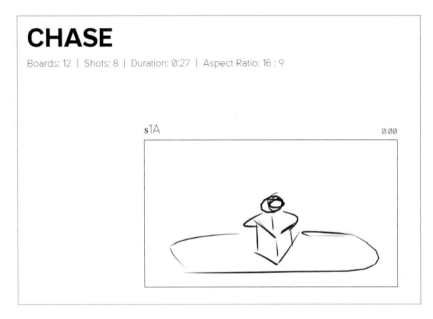

Figure 3.7 The storyboard with shot labels that were later used for filenames

The same naming convention was used to name the captured clips, so when you look at the media items in the Video Clips bin, you can easily associate each clip with the storyboard.

Of course, you can come up with your own naming convention for your productions.

Creating the sequence

The project needs at least one sequence, so create one now:

1. Look at the storyboard to find out which video clip should be first. It's s1A. However, the Video Clips bin contains filenames such as s1aV2.mp4 and s1aV3.mp4, so identify which clip should be first.

   In the filenames, V2 and V3 represent different versions of, or options for, the same shot. One is wide; the other is close up. The storyboard indicates that the sequence begins with a wide shot of the student and then cuts to the closeup, so make the wider shot appear first in the sequence.

2. Create a new sequence based on the first clip using any of the techniques you learned, such as dragging and dropping the clip on the New Item button at the bottom of the bin.

3. In the Video Clips bin, change the name of the sequence so that it isn't the same as the clip name. Video 3.7 uses the name **PSA Sequence**.

4. Move the PSA Sequence from the Video Clips bin into the Sequences bin (**Figure 3.8**).

Figure 3.8 The PSA Sequence set up in the project

This is easier to do when you can see both bins at once, either by switching the Project panel to List view or by opening one of the bins in an undocked panel.

Building the first scene of the sequence

You have all of the parts you need to build the rough cut. You have the captured clips, and a storyboard that tells you how to order them in the sequence. At this point you should be able to continue adding clips to the sequence, referring to the storyboard.

Build the rough cut up through the first scene (**Figure 3.9**), as demonstrated in Video 3.7. Your version doesn't have to be an exact match of the rough cut in the video, as long as it tells the story indicated by the storyboard: a wide shot of a student reading cuts to a closeup of the student reacting to the school bell going off, and then in a series of other shots, the student looks at his watch, jumps over the table, and runs out of the room as fast as he can.

Figure 3.9 The first scene built according to the storyboard

As you build the rough cut of the first scene, keep the following in mind:

- When you select the bell.WAV file in the Audio Files bin, trim it in the Source window, add it to the sequence, and sync it up with the video, adding markers may help line up the audio and video.
- There are multiple ways to divide a clip in the Timeline, such as clicking the clip with the Razor tool ( ), positioning the playhead at a frame and choosing Sequence > Add Edit, or pressing the Add Edit shortcut (Ctrl+K on Windows or Command+K on the Mac).
- To get rid of a gap in the sequence, you can click the gap with the Selection tool ( ) and press Shift+Delete, which is a shortcut for a ripple delete.
- Take the opportunity to practice **three-point editing**, where you mark three points (source In point, source Out point, and sequence In point) and then use Insert or Overwrite to add the clip to the sequence. The duration of the clip (the duration between the source In point and source Out point) automatically determines the sequence Out point.
- Look for opportunities to use L-cuts or J-cuts when continuing audio from an adjacent clip might tell the story better than having each clip's audio start and end with its video.
- Three options were captured for storyboard frame s1c: Clips s1cV1.mp4, s1cV2.mp4, and s1cV3.mp4. You can choose which one you think tells the story best; add one of the three to the sequence and don't use the other two.

★ *ACA Objective 4.1*

★ *ACA Objective 4.3*

▶ **Video 3.8**
Vertical Video

Editing with Vertical Video

According to the storyboard, the second scene of the sequence uses shots s3a and s3b. But the clip file s3a.mp4 was captured with a smartphone with the device held vertically. Although smartphone cameras are capable of capturing good-quality video, vertical video can be a challenge to edit because the tall, narrow frame aspect ratio is difficult to reconcile with the short, wide frame aspect ratio of typical HD video.

Ideally, contributors should be reminded to capture smartphone video with the device held horizontally. But that doesn't always happen, so this section explores ways to make vertical video work in a conventional horizontal video frame.

Adding vertical video to the sequence

The second scene of the rough cut involves a vertical video from a smartphone, clip s3a.mp4. First, add it to the sequence:

1 In the Video Clips bin, open clip s3a.mp4 into the Source Monitor.

2 Set an In point and Out point for the clip.

3 Add the clip to the sequence after the first scene (**Figure 3.10**).

Figure 3.10 Clip s3a.mp4 added to the sequence

The problem is that only the middle of the vertical video is visible. The top and bottom of the vertical video is cut off by the short, wide sequence frame, and the empty left and right sides of the sequence frame have black bars because the vertical video isn't wide enough to cover them. You can't scale up the vertical video to fill the sides or you'll see even less of the vertical video, but you can scale down the vertical video so that you can see its complete top and bottom.

4 With the clip selected, right-click (Windows) or Control-click (macOS) the clip and choose Set To Frame Size (**Figure 3.11**).

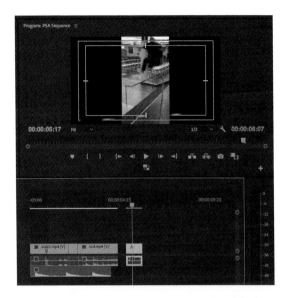

Figure 3.11 The clip set to fit within the frame

Now the entire vertical clip is visible, but the empty black areas need to be filled.

Filling the black bars with a copy of the clip

A common fix for vertical video is to fill the empty black areas with something. The simplest way to do that is with a modified copy of the same clip. Try this option:

1. Drag the video of clip s3a.mp4 to video track 2 (**Figure 3.12**).

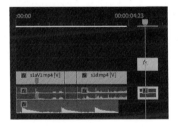

Figure 3.12 Moving the video to video track 2

2. With the same vertical video clip open in the Source Monitor, drag just the video to video track 1 in the Timeline, into the space where the original video clip used to be.

    Remember, in order to drag just the video from the Source Monitor, drag the video icon below the preview.

3. Select the copy of the vertical video on the lower video track.

4   In the Effect Controls panel, adjust the Scale value so that the clip copy covers the empty black areas.

    The fastest way to do this is to scrub the Scale value to the right (position the pointer over the value and drag to the right).

5   In the Effects panel, find the Gaussian Blur effect and apply it to the clip you scaled.

    Remember that a fast way to find an effect is to use the search field at the top of the Effects panel, and you can apply the effect by dragging it from the Effects panel and dropping it on the clip in the Timeline panel.

6   In the Effect Controls panel, increase the Blurriness value for the Gaussian Blur effect until the selected clip blurs so that it becomes a complementary background that lets the viewer focus on the vertical clip (**Figure 3.13**).

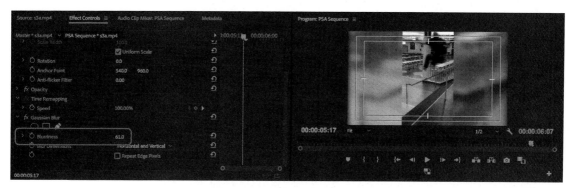

Figure 3.13 Adjusting the Gaussian Blur effect in the Effect Controls panel to blur the background

Filling the black bars with a graphic

If you have time for more preparation, another way to hide the black areas is to create a graphic background that the vertical video fits within. For this exercise, a graphic is already prepared, so you can try this option right now:

1   In the Project panel, open the Graphics bin.

2· Drag the file stairs-graphic.jpg from the Graphics bin and drop it on the blurred video copy you just created so that the graphic replaces it (**Figure 3.14**); edit the duration of the graphic so that it's the same duration as the video on the upper track.

The graphic is a scaled version of the stairs from the video along with two more elements: a red arrow illustrating the jump, and a template of a smartphone with a blank screen. It's intended to let you position the vertical video on the smartphone screen, which you can easily do with the Effect Controls panel.

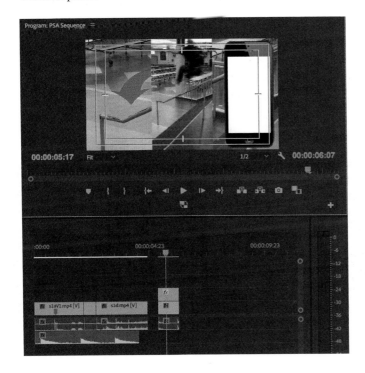

Figure 3.14 Graphic added to the sequence

3. In the Effect Controls panel, with the vertical video clip selected in the Timeline, adjust the Position and Scale values until the smartphone video fits nicely into the blank smartphone screen (**Figure 3.15**).

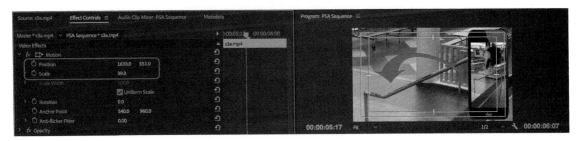

Figure 3.15 Using Position and Scale to fit the vertical video on the smartphone template's empty screen

Video 3.8 scales and positions the vertical video into the smartphone using the Corner Pin effect. The Corner Pin effect can be better when the video doesn't line up exactly with the template, as in a case where the template is slightly askew. If you want to try this, you can apply the Corner Pin effect to the selected video as you do any other effect, and then you simply drag to position each of the four corners of the effect's bounding box at the four corners of the space available in the template. This will adjust the geometry of the video to match the geometry of the template.

There may be times where you can use vertical video as it is, such as when the video is actually intended to play back on vertically held smartphones. For example, Instagram stories show video vertically.

Adding the landing to the stair sequence

In the Video Clips bin, clip s3b.mp4 is of the student landing from his jump over the stairway. Set the In point and Out point for clip s3b.mp4 in the Source Monitor, and add it to the Timeline immediately after clip s3a.mp4.

As Video 3.8 demonstrates, you may want to edit scene 3 as an L-cut, extending the audio of clip s3a through clip s3b. Remember that you have to use special techniques to edit a clip's audio independently of video, such as Alt-dragging (Windows) or Option-dragging (macOS) the end of the audio clip in the Timeline, or locking the video track.

After the L-cut is complete, consider keeping the scene together as a unit by grouping it (select the scene's clips and choose Clip > Group) and color-coding the clips (choose Edit > Label). Both commands are also available if you right-click (Windows) or Control-click (macOS) a clip in the Timeline.

★ ACA Objective 4.1
★ ACA Objective 4.3
★ ACA Objective 4.7

▶ *Video 3.9*
Multicam

Editing a Multicam Sequence

It's common to use multiple cameras to record a scene so that a scene is covered by multiple focal lengths (such as a wide shot and a telephoto detail shot) or multiple angles. The editor can then choose which shots to use at any time in the scene.

To work with multiple camera shots efficiently, you'll find it helpful to be able to play back multiple shots at the same time, perfectly synchronized. Although you could do this manually by placing shots on different tracks and lining them up, that approach would take a lot of time and effort, especially if you have many multicamera scenes. Fortunately, Premiere Pro can automatically associate multiple camera clips into a single multicam scene, while also automatically synchronizing all of those clips in time.

Creating a multicam sequence

In the third scene of the sequence, clips s2a.mp4 and s2b.mp4 are two different angles of the same shot, so you can use the multicam feature to help build it.

1. In the Video Clips bin, select clips s2a.mp4 and s2b.mp4.
2. Choose Clip > Create Multi-Camera Source Sequence. The Create Multi-Camera Source Sequence dialog box appears (**Figure 3.16**).

    That command is also available if you right-click (Windows) or Control-click (macOS) the selected clips.

3. For Video Clip Name+, enter a name such as **corner** (since the character is coming around a corner). Because there is a plus sign (+) at the end of the option name, that means "corner" will be added to the end of the name of the first selected clip. You could also click Video Clip Name+, choose Custom, and enter a name; the Custom option means the name will appear by itself.

4. For Synchronize Point, choose Audio. This works well if the audio of every clip starts with a well-defined short sound spike such as a clap. Premiere Pro can use that audio spike to line up frames precisely. Whether other options work better depends on how the video was shot; for example, In points might work if you are absolutely sure that all clips start at exactly the same frame relative to the action.

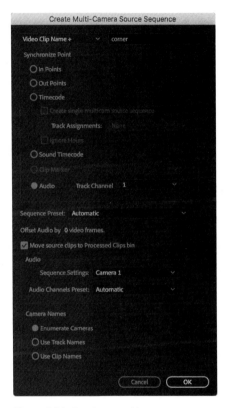

Figure 3.16 Creating a multicam sequence

5. Select Move Source Clips To Processed Bin. This will create a bin called Processed Clips (if it doesn't already exist) and will move clips s2a.mp4 and s2b.mp4 into it. That way, you'll know those two clips are already used in a multicam sequence.

6. Leave the rest of the options as they are, and click OK.

    Two new items should now be in the Video Clips bin: a sequence named s2a.mp4corner and a new Processed Clips bin containing clips s2a.mp4 and s2b.mp4.

Adding the multicam sequence to the Timeline

With the multicam sequence created, you can now use it in your main sequence:

1. Drag the s2a.mp4corner sequence from the Video Clips bin, and drop it after the end of the sequence you're working on.

   When you play back the sequence, you always see the view of just one of the cameras in the multicam sequence. But which camera? If you look at the multicam sequence in the Timeline, its name starts with [MC1]. This stands for Multicam 1, or the first camera.

   You can instantly change which camera you see.

2. With the multicam sequence selected in the Timeline, choose Clip > Multi-camera > Camera 2.

   In practice, you'll probably find it faster to switch cameras by right-clicking (Windows) or Control-clicking (macOS) the multicam sequence and choosing Multi-camera > Camera 2.

This is just the beginning of what you can do with a multicam sequence, so now that you've added it to the Timeline, you can start taking full advantage of those multiple cameras.

Cutting between cameras

A major reason to create a multicam sequence is to be able to cut between cameras quickly and easily. This is made easier if you first add a control to the Program panel controller.

1. In the Program panel, click the Button Editor button () and drag the Toggle Multi-Camera View button onto the controller (**Figure 3.17**).

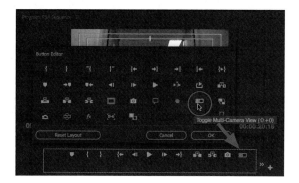

Figure 3.17 Adding the Toggle Multi-Camera View button to the controller

2. Click the Toggle Multi-Camera View button to enable it.

   The Program panel splits to show a view of all available camera clips on the left and the actual sequence on the right (**Figure 3.18**). In the view on the left, the clip with the yellow border is the one currently visible in the sequence.

Figure 3.18
Multi-Camera View in use

3. Switch to the camera view that you want to see at the start of the multicam sequence. For this project, it will be best to start with Camera 1.

4. Position the playhead at the beginning of the multicam sequence and begin playback, watching for a time when you'd like to cut to another camera.

   For more precision, you can position the playhead frame by frame instead of playing back the sequence.

5. With the sequence playing, when you see the moment you want to cut to another camera, click that camera in the Multi-Camera View. The view you clicked now has a red border.

6. Stop playback (for example, press the spacebar).

   Two things happen: The sequence display on the right side of the Program panel switches over to the other camera, because Premiere Pro has also created an edit point and switched cameras in the video track (**Figure 3.19**). The red border in the Multi-Camera view changes back to a yellow border to let you know that playback has stopped and the edit has been applied to the sequence.

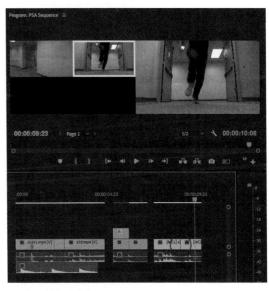

Figure 3.19 The Edit point in Timeline video track created by clicking in multicam view

Chapter 3 Editing an Action Scene   175

Using this method, you can simply watch the multicam sequence and click the multicam display to create camera switch edits without stopping playback. This lets you edit a multicam sequence on the fly, almost as if you were switching between cameras during a live program.

7   As needed, refine the timing of the edits within the multicam sequence using any of the other editing techniques you have learned, such as rolling, and slip/slide edits.

8   Group and label this scene, and save the project.

Practicing multicamera edits

> **TIP**
> When using multiple cameras to record a performance, try to record the entire program at once, or at least with as few breaks as possible, to minimize the number of multicam sequences you have to create and synchronize.

The fourth sequence is represented in the Video Clips bin by clips s4a.mp4 and s4b.mp4, which is another set of multicamera shots of the same scene. Use the skills you just learned to build a multicamera sequence from those two clips, add that to the end of the main sequence, find the right point to cut between cameras, and refine the edits.

You're done with multicamera edits, so click the Toggle Multi-Camera View button in the Program panel to switch back to the normal Program panel view.

ACA Objective 4.1

ACA Objective 4.3

Video 3.10
Labeling

Finishing Sequence Edits

There are just a few more things to do before all clips are in place. You need to add the last clip and clean up gaps.

According to the storyboard, the last clip should be s4c, where the student is relieved that his good fitness level allowed him to arrive on time to class.

To add the last clip, drag clip s4c.mp4 from the Video Clips bin to the end of the sequence, and trim as needed.

If you haven't already removed the gaps between scenes, this is a good time to do that. Earlier you learned how you can select a gap and press Shift+Delete to perform a ripple delete. That can become tedious if you have many gaps to remove, but there's a faster way.

To remove all gaps in a sequence and close the gaps, choose Sequence > Close Gap (**Figure 3.20**).

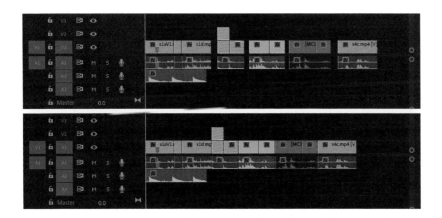

Figure 3.20 Before and after using the Close Gap command

Another editing detail you can add is to have the school bell sound again at the end of the sequence. Copying the audio clip at the beginning of the sequence and pasting it at the end may add it on the wrong track, but by changing which tracks are targeted for audio, you can make it paste on the intended track, Audio 1.

At this time, you may want to make sure that each scene is labeled with a distinctive color so that in the future if you or someone else needs to go in and make changes, it's obvious where each scene starts and stops.

You may be done with the clip edits, but there's still some polishing to do to the clip audio and video.

> **TIP**
> *Another quick way to duplicate the school bell is to Alt-drag (Windows) or Option-drag (macOS) to create a copy of it and drop it at the end of the Timeline.*

Sweetening Different Audio Types

★ ACA Objective 4.7

 *Video 3.11 Essential Audio*

Professional video productions rarely use audio as it is. Audio is often processed and refined, similar to how you might color-correct and grade video clips, to create a more cohesive overall program that also reinforces the story and the director's intentions.

Setting up for audio editing

Before editing audio, you'll need to set up your workspace for audio editing as you have before.

1 To begin, do one or more of the following as needed:
- Switch to the Audio workspace.
- In the Timeline panel, adjust the height of the audio tracks so that waveforms and controls are easier to see.

- Next you'll add the background music you were asked to find earlier in the chapter. Because the sequence is only around 15 seconds long, your music file is probably much longer, so you'll probably have to trim it down.
- If you've been following the recommendations for project organization, you should have copied your background music file into the Audio Clips folder so that it's with the other project files.
- If your music clip is not much longer than the sequence, you can skip step 2 and trim it in the Timeline panel instead.

2   Double-click your background music file to open it in the Source Monitor, find the best 20 seconds of it to use for the sequence, and set In and Out points for it.

You want the music to be a few extra seconds longer than the sequence video to allow for end credits you'll add later.

3   Drag the background music clip into an unused audio track in the Timeline panel (**Figure 3.21**).

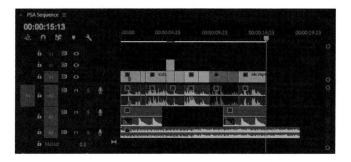

Figure 3.21 Background music audio clip added to audio track 3

Sweetening audio using Essential Sound

In conventional audio sweetening, audio engineers use their expertise to adjust audio using specialized audio controls, based on the type of sound being adjusted. In Premiere Pro, the Essential Sound panel automates this process so that you can treat audio by indicating what kind of audio it is.

To sweeten clip audio using Essential Sound:

1   In the Timeline, select the first and last clips in the sequence, which both contain dialogue.

2   If the Essential Sound panel isn't visible, choose Window > Essential Sound.

3   In the Essential Sound panel, click Dialogue. This tells Essential Sound that the audio in the clips you selected should be treated as dialogue, and dialogue controls appear in the Essential Sound panel (**Figure 3.22**).

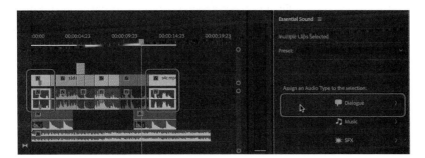

Figure 3.22 After selecting the dialogue clips in the Timeline, click the Dialogue button in the Essential Sound panel.

4   In the Timeline, select all the video clips except the first and last clips; you can do this by clicking the second clip and Shift-clicking the second-to-last clip.

5   In the Essential Sound panel, click Ambience. This tells Essential Sound that the audio in the clips you selected should be treated as ambient sound, and appropriate controls appear in the Essential Sound panel.

6   In the Timeline, select the two instances of the bell audio clip, and in the Essential Sound panel, click SFX (sound effects).

7   In the Timeline, select the background music clip, and in the Essential Sound panel, click Music.

8   In the Essential Sound panel, click the Preset menu and choose Balanced Background Music (**Figure 3.23**). This preset helps to balance the music with the other types of sounds in the other clips.

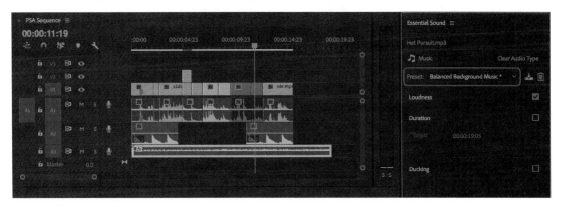

Figure 3.23 Choosing the Balanced Background Music preset

Chapter 3 Editing an Action Scene   **179**

9 If you want to fade out the background music, add keyframes to the audio level rubber band for the background music clip, as you learned earlier.

10 Play back the sequence to hear how it sounds. Feel free to adjust the Essential Sounds settings for any of the clips to further refine the sound mix.

★ *ACA Objective 4.5*

▶ ***Video 3.12***
Adjustment Layers

Applying an Adjustment Layer

Suppose you want to add an effect to an entire sequence. You already know that you could simply apply the effect you want to one clip and then use the Edit > Copy and Edit > Paste Attributes commands to paste effects to other clips. But that is not always practical for complex, multiple-track sequences.

There is a better way: using an **adjustment layer**. An adjustment layer is like an empty media item that does nothing but apply effects to all tracks below it. For example, instead of pasting a color effect onto many clips in a sequence, you can add one adjustment layer to a track above the video tracks and apply the color effect to just the adjustment layer, and the color effect on the adjustment layer will affect all tracks under it.

If you have used adjustment layers in Adobe Photoshop or Adobe After Effects, you already know how they work.

To use an adjustment layer:

1 In the Project panel or a bin, click the New Item button and choose Adjustment Layer.

You can also create an adjustment layer the same way you can create the other items on the New Item button menu: choose File > New, or right-click (Windows) or Control-click (macOS) in the Project panel or a bin and choose New Item.

2 In the Adjustment Layer dialog box, make sure the settings are appropriate for the sequence you want to add it to, and click OK.

The new adjustment layer appears in the Project panel or bin. You can rename it the same way you rename any other item in the Project panel or bin: by clicking to highlight it. Renaming is recommended if you'll be using more than one adjustment layer so that you can tell them apart.

3 Drag the adjustment layer from the Project panel or its bin and drop it on a Timeline track above other video tracks (**Figure 3.24**).

If there are no unused tracks above existing tracks, simply drop the adjustment layer above existing tracks, and Premiere Pro will automatically add a new track for the item you drop.

Figure 3.24 A new adjustment layer in the Project panel and added to the Timeline

4 Drag either end of the adjustment layer so that its duration is as long as the sequence, or however long you need it to be.

For this sequence, have the adjustment layer end where the last video clip ends so that it doesn't affect the credits you'll add later.

5 Apply an effect to the adjustment layer (**Figure 3.25**). As you play back the sequence, you'll notice that the effect applied to the adjustment layer affects all clips in lower tracks.

Figure 3.25 An adjustment layer applying a Lumetri Color look across the entire sequence

The example demonstrated in Video 3.12 shows a Lumetri Color effect being applied to an adjustment layer—specifically, a look from the Creative group of settings. A look is a color style that's applied to simulate a type of film or evoke a mood. In a longer sequence, you might have adjustment layers that cover only specific scenes in the sequence—for example, as a sequence changes from warmly lit sunset scenes to cool bluish night scenes and sepia-toned scenes from one character's memories.

The advantage of using an adjustment layer to apply an effect is not just that it's easier to apply to many clips. An adjustment layer also makes it much easier to edit an effect applied to many clips, because instead of having to go back and edit numerous clips, you edit the effect just once on the adjustment layer and it changes all the clips that it affects, no matter how many there are. That's a major time savings.

ACA Objective 2.2

ACA Objective 2.3

▶ **Video 3.13**
Timeline Interface

Reviewing Timeline Controls

The work you have done so far has given you experience and practice in the fundamentals of Premiere Pro. At this point it's a good time to review some Timeline productivity tips that you may need to know if you haven't already discovered them.

The group of controls near the top-left corner of the Timeline (**Figure 3.26**) let you refine to how you use the Timeline:

- **Playhead Position time indicator:** This isn't just for show. You can click it to type the sequence time you want to go to, and you don't even have to include punctuation. For example, to go to 10 seconds 0 frames, type **1000** and press Enter. Premiere Pro will interpret 1000 as 10:00. Also remember that you can scrub the Playhead Position time indicator as an alternative to dragging the playhead along the Timeline.

- **Insert And Overwrite Sequences As Nests Or Individual Clips.** When you add a sequence to another sequence, do you want the sequence to be added as one media item under the sequence name, or do you want each of the clips in the sequence to be added individually? If you want the former, leave the button at its default setting, selected. If you want the latter, click to deselect the button. (You'll work with nested sequences later in this chapter.)

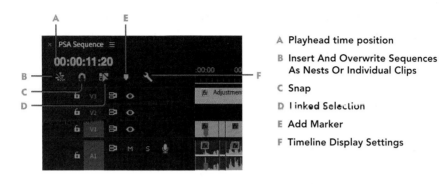

Figure 3.26 Timeline control group near the playhead time position display

A Playhead time position
B Insert And Overwrite Sequences As Nests Or Individual Clips
C Snap
D Linked Selection
E Add Marker
F Timeline Display Settings

- **Snap.** When selected, the Snap button enables snapping edges of Timeline items to each other, such as clips and markers. Press the S key to enable or disable snapping with a keyboard shortcut.

- **Linked Selection.** When selected, the Linked Selection button selects both the video and audio of a clip if you click either. You might turn this off when you want to edit clip audio separate from video, such as working with L-cuts or J-cuts for the next hour. When you're done, you might want to enable the Linked Selection button again.

> **TIP**
>
> Instead of using the Linked Selection button, you can toggle that setting by pressing the S key, even while editing in the Timeline.

- **Add Marker.** As you learned earlier, clicking the Add Marker button inserts a marker into a sequence or a selected clip. The shortcut for this button is pressing the M key.

- **Timeline Display Settings.** Click the wrench icon () to customize the display of the Timeline. For example if you aren't seeing fx badges on clips, you may need to enable them here.

You'll find additional display options in the Timeline panel menu, such as whether thumbnails are displayed in the Timeline for clips.

You may also want to review other Timeline panel controls in "Navigating the Timeline," in Chapter 1.

Adding Credits

One of the finishing touches of a video program is to add credits and other titles. You've already created titles during this course, but this time you'll add a new twist: rolling credits.

★ ACA Objective 4.2

 *Video 3.14* Rolling Credits

Adding credits using a template

You'll once again use the Essential Graphics panel to streamline the process of making titles.

To add credits:

1. Open the Essential Graphics panel (Window > Essential Graphics).

   Alternatively, you can switch to the Graphics workspace, which should open the Essential Graphics panel.

2. In the Essential Graphics panel, make sure the Browse tab is active, drag a title template that you'd like to use, and drop it on a video track after the last video clip, over the music at the end of the sequence. Video 3.14 uses the Bold Title template (**Figure 3.27**).

Figure 3.27 Dragging the Bold Title template into the sequence

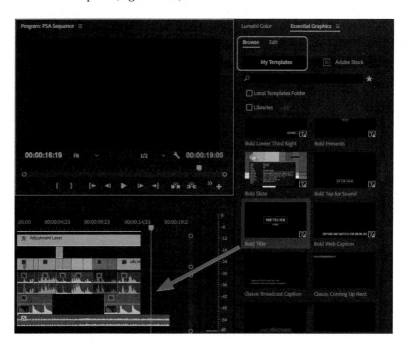

3. If the Resolve Fonts dialog box appears, select any fonts that need resolving, and click OK.

   If your Creative Cloud account does not have access to Adobe Typekit, it may not be possible to resolve fonts. For this chapter, that's okay; simply click Cancel. At any time, you can change the font to one that's installed on your system.

4. Play or scrub the duration of the title to preview the template. The graphic and text of the title animate from a small size to full size.

5. In the Program panel, double-click the text in the title template and replace it with the text you want to use. Video 3.14 uses **Stay in Shape!** for the title and **Don't be Tardy** for the subtitle (**Figure 3.28**).

Figure 3.28 The edited title

Adding rolling credits

Next, you'll add rolling credits. The text file demonstrated in Video 3.14 is not supplied with the chapter files because it's specific to the music clip used in the video. Because this chapter directed you to find a music clip to use, you'll need to write your own credits and any necessary attributions for the music you found. The rolling credits you create here don't involve a template.

To add rolling credits:

1. In a text editor, prepare a text file with all the credits you need to show.

   It's much easier to write and edit a long list of credits in a text editor than in the Program panel.

2. In the text editor, select all the credits text.

3. Switch to Premiere Pro, and in the Timeline panel, move the playhead to the time when you want the credits to start.

4. Select the Type tool (T), and click in the Program panel.

5. Choose Edit > Paste or press its keyboard shortcut, Ctrl+V (Windows) or Command+V (macOS).

   The text on the clipboard is pasted into the text layer that you created by clicking with the Type tool (**Figure 3.29**). But it probably needs formatting.

Figure 3.29 Credits text pasted into the text layer created by clicking the Type tool

6. With the text layer selected, use any text controls in the Essential Graphics panel to format the characters and paragraphs so that the credits are an appropriate and legible size and format for all of the screen sizes on which you expect viewers to watch the video.

7. Deselect the text layer by clicking in an empty area of the Program panel or text layer list in the Essential Graphics panel, but leave the title clip selected in the Timeline.

   When a title is selected in the Timeline but no text layers are selected, the Edit tab of the Essential Graphics displays options for the entire clip, including the Roll option.

   When the Roll option is enabled, a scroll bar appears in the Program panel so you can preview the roll settings. However, you can preview roll timing only by playing back the sequence.

8. Select the Roll option, and adjust the Roll options as needed (**Figure 3.30**):

   Preroll and Postroll add extra time before and after the credits, respectively, in case you want to delay the start or leave time after the end of the title's duration.

   Ease In and Ease Out are forms of easing, or ramping the credit roll speed up or down, respectively. If you think the credits start or stop rolling too abruptly, add Ease In time to start motion more gradually or add Ease Out time to make the credits decelerate to a stop.

Figure 3.30 Dragging the scroll bar in the Program panel to test rolling credit settings

TIP

To change the speed of rolling credits, change the duration of the credits item in the Timeline. A longer duration slows the rolling credits. The more lines of text there are, the more time you should allocate to the credits.

9. Play or scrub the duration of the title to preview the template. The credit will roll onto the screen according to the settings you specified.

Recording a Voiceover

★ ACA Objective 4.7

▶ **Video 3.15** *Voice Recording*

Voiceover audio is so common that Premiere Pro provides a way for you to record it directly into a project.

Voiceover recording starts in preproduction, where you should write, edit, and proof your voiceover scripts in advance. In the tutorial video, the voiceover is simple: just the line "Stay in shape, don't be tardy" to reinforce the closing title you added earlier.

Don't underestimate the value of audio; it can be as important as video in giving viewers the impression of quality in your production. If you'll be recording voiceovers frequently or for high-value clients, you'll want to research how to make better recordings in your studio. Though it can start with buying a good-quality microphone for voice, how your studio is set up also makes a big difference. For example, speaking close to the microphone and installing acoustic treatments to reduce room reflections can make your recordings sound professional. Here's the general workflow for recording a voiceover. You'll begin by preparing for recording.

Preparing Audio Hardware preferences for audio recording

Before recording, it's a good idea to make sure Premiere Pro is properly configured to record and play back audio. At the very least, check the following settings:

1. Choose Edit > Preferences > Audio Hardware (Windows) or Premiere Pro CC > Preferences > Audio Hardware (macOS). The Preferences dialog box opens to the Audio Hardware pane (**Figure 3.31**).

2. Make sure that Default Input is set to the microphone or audio interface you are using, especially if you've connected a microphone.

3. Make sure that Default Output is set to the speakers, audio output port, or audio interface you are using, especially if you're using separate speakers.

4. If you're using speakers to monitor audio, click Audio in the list on the left to open the Audio pane, and deselect Mute Input During Timeline Recording. This will help prevent feedback when you record.

   You can leave that option deselected if you are monitoring through headphones.

5. Click OK.

Figure 3.31 The Audio Hardware preferences pane

You should recheck the Audio Hardware settings if Premiere Pro isn't capturing audio that you're recording or if you can't hear audio that should be in a clip or sequence, especially if you've changed the audio equipment connected to your computer.

Preparing the Timeline for audio recording

The Timeline provides some controls that are useful for audio recording, but they aren't displayed by default, so you can make them visible. If you don't see the Voice-Over Record button ( ) before each audio track, you can add it using a button editor that's similar to the one you used in the Program panel.

1 Click the Timeline Display Settings icon (**Figure 3.32**) and choose Customize Audio Header.

   You can also right-click (Windows) or Control-click (macOS) the track header area to the left of the audio tracks and choose Customize.

2. In the Button Editor, drag the Voice-Over Record button into the audio track header area, and click OK.

3. Right-click (Windows) or Control-click (macOS) the track header area to the left of the audio tracks and choose Voice-Over Record Settings (**Figure 3.33**). In this dialog box you can adjust recording settings, including the following:

   - **Name:** The name of the clip that will be saved after you record
   - **Source:** The audio source hardware
   - **Input:** The audio input enabled from the source
   - **Countdown Sound Cues:** Whether you want to hear sounds that prepare you for when the recording starts
   - **Preroll and postroll:** As with titles, these add extra time before and after, respectively, to the actual recording time.

   There is also an audio level meter that you can use to make sure your microphone signal is strong enough to record well. As you speak into the microphone, the green bars should extend up to about –6dB without peaking in the red area. If you don't see green bars, make sure your audio source and input are connected properly and enabled. If the green bars extend into the red zone, reduce the audio level coming into the computer, or reduce the input audio level in your operating system.

4. Click Close.

Figure 3.32 Clicking the Timeline Display Settings icon to choose Customize Audio Header

Figure 3.33 Voice-Over Record Settings dialog box

Recording voiceover audio

With Premiere Pro set up to record audio, you can go ahead and record the voiceover.

1. In the Timeline, position the playhead at the time where you want to start recording.

2. In the Timeline, click the Voice-Over Record button next to the audio track where you want the recording to be added (**Figure 3.34**).

Chapter 3 Editing an Action Scene **189**

Figure 3.34 The Voice-Over Record button next to an audio track

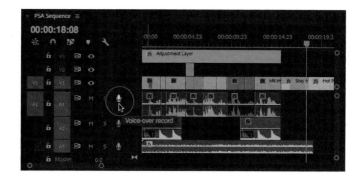

TIP

If a voiceover audio clip is too loud or too quiet, you can adjust its level the same way you would for any other audio clip in the Timeline: drag its audio level rubber band up or down, or choose the Clip > Audio Options > Audio Gain command when the clip is selected.

3. Speak the voiceover into the microphone at the time that it needs to be heard. Your timing doesn't have to be exact because you can edit the timing of the audio later.

4. When you're finished recording, stop recording by clicking the Voice-Over Record button or pressing the spacebar.

5. Play back the part of the Timeline where you added the recording to make sure it's good.

6. In the Project panel, locate the new audio file, and drag it to the Audio Clips bin if it isn't already there.

TIP

If you can't find the new audio file created by voice-over recording, right-click (Windows) or Control-click (macOS) the clip in the Timeline and choose Reveal In Project. The clip will become selected in the Project window or its bin.

7. Save the project.

Nesting Sequences for Different Delivery Requirements

★ ACA Objective 3.1

▶ Video 3.16
Working with Nesting and Special Sequences

There are more ways than ever to deliver video, and unfortunately, they have different requirements. For example, whereas 16:9 aspect ratio video is standard for television, many movies are composed for other aspect ratios, and some social media channels work best with 1:1 (square) aspect ratio video.

When you need to adapt a video for different specifications, you don't have to start over and build the sequence again. You can copy the sequence and change only the parts that need to be adapted, and you can also **nest** one sequence inside another and have the containing sequence use different specifications. Nesting means using a sequence as a media item in another sequence, as if the first sequence was a clip.

Video 3.15 discusses preparing this project's 16:9 aspect ratio sequence for use as a 1:1 video for posting on Instagram.

Creating new sequences for social media

In some cases, you can just drag one sequence into another to nest them. However, in this case, there will need to be some changes to the sequence to adapt it to a square aspect ratio and you don't want to alter the finished original sequence. For this reason, the finished sequence will first be duplicated, and the adaptations will be made only to the duplicate.

1. In the Sequences bin in the Project panel, select the sequence you created in this chapter and choose Edit > Duplicate.
2. Click the duplicate sequence and rename it **Social Media Draft**.
3. Create a new sequence using any of the methods you've learned so far.
4. In the New Sequence dialog box, make sure the Sequence Presets tab is selected, expand the Digital SLR preset group, expand the 1080p preset group, and select DSLR 1080p24, which most closely matches the specifications of the sequence you created.
5. Click the Settings tab and change the frame size for both horizontal and vertical to **600**.
6. For Sequence Name, enter **Social Square**, and click OK.
7. If the Social Square sequence isn't already added to the Sequences bin, move it there.

Nesting one sequence inside another

Now you're ready to nest sequences.

1. In the Sequences bin, drag the Social Media Draft sequence into the Social Square sequence.
2. Double-click the Social Square sequence to open it in the Program and Timeline panels.
3. Drag the Social Media Draft sequence from the Sequences bin, and drop it at the beginning of the Social Square sequence in the Timeline panel. If the Clip Mismatch Warning alert appears, click Keep Existing Settings.

The Social Media Draft sequence is now nested inside the Social Square sequence, like a clip. But it isn't sized to fit the 600x600-pixel frame, so you'll need to scale it.

4 Select the Social Media Draft sequence in the Timeline, and in the Effect Controls panel, adjust the Scale percentage until it vertically fills the square frame without black bars (**Figure 3.35**).

Figure 3.35 The Social Media Draft sequence nested as a clip in the Social Square sequence and scaled to fit

The sides of the 16:9 wide aspect ratio sequence will not be visible in this 1:1 square aspect ratio sequence, but that's expected. Soon you'll make adjustments for that.

TIP

Remember that a quick, interactive way to resize with the Effect Controls panel is to simply scrub the Scale value.

5 Play back or scrub the sequence to see how the wide video plays in this square frame, and make notes on what parts of the video are not showing up properly within the square frame.

6 In the Timeline panel, click the tab for the Social Media Draft sequence and perform the adjustments needed to make the video work in the square frame, such as repositioning or recomposing video, graphics, or credits.

7 Check your work by clicking the tab for the Social Square sequence and playing it back.

One feature of nesting is that any changes you make to nested sequences will appear in the containing sequence. Some editors use this feature to help organize very long programs. They edit each act or major scene as its own sequence of clips, and then nest those inside a master sequence.

When you're done, you're ready to export the Social Square sequence for social media. You won't need to export the Social Media Draft sequence, but you'll need to keep it around since it provides the content for the Social Square sequence.

Using Proxies and Removing Unused Clips

★ ACA Objective 2.1
★ ACA Objective 2.4
★ ACA Objective 3.1

▶ Video 3.17 *Proxy and Unused Clips*

As you learned in Chapter 1, video editing demands high performance from every part of a computer system, especially the CPU, the graphics hardware, and the storage drives. It doesn't take long for a project to become complex enough that an average computer might start to struggle to smoothly play back a sequence without taking the time to render previews first. These issues are magnified when editing 4K video sequences, which can strain even expensive high-performance computers.

Premiere Pro can create proxies (low-bitrate copies) of clips so that a sequence can play back more smoothly. As you edit, you can use the Program panel to switch to proxies for faster editing, or switch to the original clips for maximum image quality.

A proxy can be useful any time the data rate is too high relative to the processing ability of the computer. Though proxies are popular for editing 4K video on powerful computers, proxies can be just as useful editing 2K (1080p) video on less powerful computers.

When you export, Premiere Pro will use the original full-quality clips, not the proxies.

ABOUT PROXY WORKFLOWS

You may ask why a video editing program would have difficulty playing 2K (1080p) or 4K video when a simple media player might have no problem playing back those resolutions on a smartphone or on a streaming box for a television. The answer is that it is not difficult to simply play back one finished video file, but when a clip is in an editing timeline with other clips, there is significant additional processing involved in keeping clips ready for editing—especially when you start piling on tracks, transitions, and effects that add to the calculations needed to display each frame.

One useful solution is the *proxy* workflow. A proxy is a version of a clip that is a lot less demanding for a computer to play back. Technically, a proxy has a much lower data rate or bitrate than the original, and that's what makes it faster and easier to use in a timeline. A proxy may also use a codec that doesn't stress the processor as much during editing. Premiere Pro makes it easy to generate proxies and switch to them during editing.

continues on next page

continued from previous page

There is a trade-off: the much lower data rate means that a proxy has lower image quality than the original. But that's usually okay for sequencing clips and timing edits and transitions. When it's time to use the originals for detail-oriented work or critical color correction and grading, it's easy to switch from a proxy to an original clip in Premiere Pro.

When a project bogs down and won't play back smoothly, and you know you've done the best you can to get the most performance out of your system, using proxies may make it possible to edit with less lag and maybe even play back a sequence in real time.

Setting up for proxies

Proxies should be subject to the same good file management practices that you use for all other media files used in the project. You should know where they are stored in case you have to relink to them and because you may want to discard them after you're done editing the project.

For this project, store the proxies in a folder next to the original clips:

1 In the Project panel, open the 4K Clip bin and create a new bin in it.

2 Name the bin **Proxy Clips**.

There's also one control to add in the Source and Program panels using the Button Editor, which you've used before:

1 In the Source Monitor, click the Button Editor button.

2 Drag the Toggle Proxies button to the controller (**Figure 3.36**), and click OK.

Figure 3.36 Adding the Toggle Proxies button to the controller

This exercise involves only the Source Monitor, but if you plan to also use proxies in sequences, repeat steps 1 and 2 in the Program Monitor so that the Toggle Proxies button will also be available there.

Creating proxies

You can create proxies as you import clips or for clips that are already in the project. In this case we're creating a proxy for a clip that's already in the project.

To create a proxy from a clip in the project:

1. Select the clip in the Project panel; Video 15.5 uses the clip DJI_0012.MOV in the 4K Clips bin.
2. Right-click (Windows) or Control-click (macOS) the clip, and choose Proxy > Create Proxies.
3. In the Create Proxies dialog box (**Figure 3.37**), choose a format from the Format menu. This determines which presets are available.

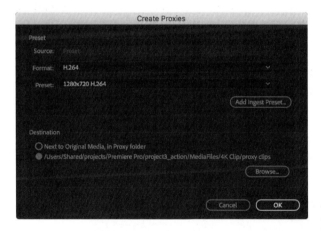

Figure 3.37 The Create Proxies dialog box

4. Choose a preset from the Preset menu. In general, smaller frame sizes produce smaller, faster proxies.
5. For Destination, click Browse, navigate to the Proxy Clips folder you created earlier, and click Select Folder.
6. Click OK. Premiere Pro sends the clip and the Create Proxies settings to Adobe Media Encoder, which renders the proxies in the background. However, it's recommended that you switch to Adobe Media Encoder so you can verify that the proxies have finished rendering before you continue to work in Premiere Pro.

You can create proxies as you import clips; select them in the Media Browser, enable the Ingest option, click the Ingest Settings button, click the Ingest Settings tab, and select the Ingest option, and choose Create Proxies (or Copy And Create Proxies) from the menu next to the Ingest option. At that point, the process is similar to creating proxies from clips in the project; you can choose a preset and a destination.

Using proxies

The proxies you create are automatically linked to the project and to their original clips, so it's easy to switch between them using the proxies and originals.

To use the proxy for the sample clip in the project:

1. In the 4K Clip bin, double-click the clip DJI_0012.MOV to open it in the Source Monitor.
2. Play back the clip in the Source Monitor. If the clip stutters or does not appear to move, Premiere Pro may have trouble playing 4K clips on your computer system.
3. Click the Toggle Proxies button (**Figure 3.38**) you added earlier to the controller.

Figure 3.38 The highlighted Toggle Proxies button means a proxy is being used.

NOTE

If a proxy has black bars or is distorted, the proxy preset has an aspect ratio that doesn't match the original clip. For best results, delete the proxy and generate it again with a different preset that has the same aspect ratio as the original clip.

4. When the button is highlighted, proxies are active and playback should be smoother. When the button is not highlighted, playback will use full-quality original clips but will be less smooth.

Creating proxies does not automatically result in smooth editing and playback. It's still possible that the combined load of processing proxies, tracks, transitions, and effects may be too much for smooth playback, especially on a less powerful computer. If this happens, experiment with different proxy formats and presets until you find one that works best on your system. Less powerful computers may require proxies that use smaller frame sizes, lower bitrates, and less demanding codecs.

You can create your own custom proxy presets using Adobe Media Encoder. For more information, see the following site:

https://helpx.adobe.com/premiere-pro/kb/ingest-proxy-workflow-premiere-pro-cc-2015.html#CreateanIngestPreset

Cleaning up unused clips

It's typical to import many captured clips that never get used in a project, especially when many options were shot for scenes for editing flexibility. When you're done editing a project, you can clear them all out of a project by choosing Edit > Remove Unused Clips.

Remove Unused Clips removes clips only from a project. The clip files remain on your storage drive.

Exporting Multiple Sequences

It's common for a job to require multiple forms of a sequence. One might be optimized for high-definition television, another for viewing on a web page, and another for social media on mobile devices. Premiere Pro and Adobe Media Encoder make it easy to export multiple sequences to different specifications.

▶ **Video 3.18**
Exporting Media Encoder Importing Preset

To export multiple sequences:

1 In the Sequences bin, select the PSA sequence and the Social Square sequence.
2 Choose File > Export > Media.
3 In the Export Settings dialog box, apply a format and a preset. For these sequences, the format is H.264 and the preset is Vimeo 720HD.
4 Change any other settings as needed, and click Queue.

   Premiere Pro CC sends the sequences and their export settings to Adobe Media Encoder, and both sequences are loaded into the queue (**Figure 3.39**).

Figure 3.39 The two sequences are loaded into the Adobe Media Encoder queue with warning triangles. The first item in the queue is the proxy job completed earlier, which is why it's dimmed.

Both items display a warning. For the PSA sequence, the warning is because the frame size of the sequence (1920x1080 pixels) doesn't match the frame size of the applied preset (1280x720 pixels). This is expected and accepted, so you can ignore the warning.

For the Social Square sequence, the warning is because the frame size and aspect ratio of the sequence are different from the applied preset. If a more appropriate preset is loaded, you could apply it to the sequence in the queue, but the preset is not loaded, so you'll have to do that first.

5 In the Preset Browser panel, click the Import Presets button.

6 Navigate to the Presets folder inside the folder for this project, select Mobile square 600x600.epr, and click Open. This loads the preset into Adobe Media Encoder (**Figure 3.40**).

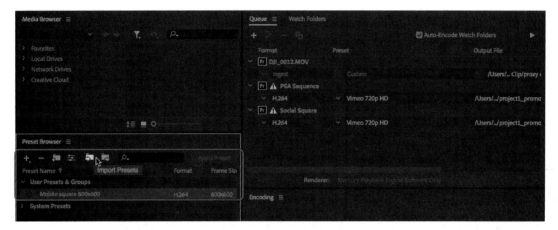

Figure 3.40 Import Presets button, and the Mobile square 600x600 preset is loaded into the User Presets & Groups set.

7 Drag the Mobile square 600x600 preset from the Preset Browser and drop it onto the Social Square sequence in the queue to apply it to that sequence.

8 Click the Output File column for each of the sequences and set the destination to the Exports folder you set up.

9 Click the Start Queue button, or press Enter or Return.

10 When the queue finishes processing, check the Exports folder to verify that the sequences exported as expected.

Using the Project Manager

★ ACA Objective 2.4

▶ Video 3.19
 Project Manager

When you're done editing a project, you'll want to archive it to a different storage drive to free up space on your working drives. The Project Manager can help you clean up a project so that you archive only what's necessary, thus saving storage space.

Remember the Remove Unused Clips command you used earlier? The Project Manager is more powerful than that. The Project Manager can remove unused clips, collect all linked media files from various folders, and put them all in one place. It can also transcode media, in case you received files in various formats and want to archive them in one consistent format.

To use the Project Manager:

1 Choose File > Project Manager (**Figure 3.41**).

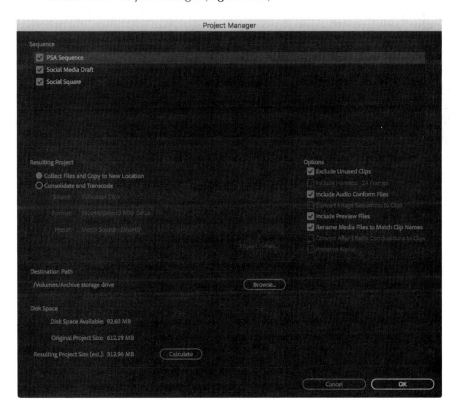

Figure 3.41 The Project Manager

2. Select the sequences you want to archive.

3. For Resulting Project, select whether you want to just collect files to one location or whether you want to consolidate and transcode (convert) to a certain format. This choice determines which options are available.

4. For Options, select the options you want to apply. If you're not sure what to do, you can accept the default settings.

5. For Destination Path, click Browse and choose the folder where you want to store the archived version of the project.

6. For Disk Space, click Calculate if you want.

   If the Resulting Project Size is larger than expected, try adjusting settings and options. For example, you might decide that you don't need to include preview files, or you might transcode to a more compact format.

7. Click OK. The Project Manager processes the project.

8. Switch to the desktop and review the new folder containing the archived project.

> **NOTE**
>
> *You may be tempted to create a ZIP document from an archived project to compress it further. Video media is already compressed, so you probably won't gain much storage space by zipping it.*

Challenge

Create your own multiple-camera sequence. It can be a fast-paced story like the one in this chapter, such as a person rushing to be on time for an appointment. Or you could create a chase scene that cuts between two people running. You can also create a how-to video that requires multiple cameras, or record a performance from multiple angles.

▶ *Video 3.20
Challenge*

As you plan your project, remember Joe Dockery's Keys to Success from the video:

1. Keep it short. Don't forget to use preproduction tools like a storyboard and shot list to maintain limits.
2. Plan out your story and include at least a couple of action sequences.
3. Possibly shoot with two cameras so that you can use the multicam feature in Premiere Pro. Remember that a smartphone can be a second camera.
4. Find music that fits the topic or make your own.
5. Share it with the world.

CHAPTER OBJECTIVES

Chapter Learning Objectives

- Apply special effects.
- Draw a simple opacity mask.
- Use the Ultra Key effect to remove a green screen.
- Animate effects with keyframes.
- Add video layers.

Chapter ACA Objectives

For full descriptions of the objectives, see the table on pages 279–283.

DOMAIN 1.0
WORKING IN THE VIDEO INDUSTRY
1.1, 1.2

DOMAIN 2.0
PROJECT SETUP AND INTERFACE
2.1, 2.3

DOMAIN 4.0
CREATE AND MODIFY VISUAL ELEMENTS
4.2, 4.5, 4.6

CHAPTER 4

Compositing with Green Screen Effects

In this project's scenario, our editor for Brain Buffet TV is out today, so we need you to edit the weather report. This will give you some great practice compositing the footage shot on our green screen with the weather graphics. You'll learn how to "key out" a background, import a layered Adobe Photoshop CC file, and light for green screen. The entire scene is only about 25 seconds long.

Preproduction

As you've learned, production can't start until the project requirements are clearly understood. Let's review them before you begin:

- **Client:** Brain Buffet TV
- **Target Audience:** Brain Buffet TV is broadcast at the Happy Old Retirement Home, so your target audience is 70–90 years old, mostly female.
- **Purpose:** The purpose of the weather report is to let the people living in the retirement home know what type of weather to expect if they go outside.
- **Deliverables:** The client expects a 20-to-30-second video featuring the weather report layered over the map and motion graphics illustrating the weather-related facts. The video should be delivered in H.264 720p. The client also requires an audio file that can be used to create a written transcript for the deaf. The audio should be delivered in MP3 format with a bitrate of 128 Kbps.

★ *ACA Objective 1.1*

★ *ACA Objective 1.2*

▶ **Video 4.1**
Introducing the Weather Report Project

Listing the available media files

Some media has already been acquired for the project. What do you have to work with? Unzip the project files using the same techniques you used for the previous project, and look through the unzipped folders:

- A master video clip shot in front of a green screen
- A weather map still image in Photoshop format with separate layers for sunshine, temperatures, and thunder and lightning
- A TV station logo still image in Photoshop format
- A hiking photo

With these items, you're ready to start setting up the project.

★ ACA Objective 2.1

★ ACA Objective 2.4

★ ACA Objective 4.5

▶ **Video 4.2**
Organize Your Project

Setting Up a Project

Start the editing stage of production by practicing the project setup techniques you learned earlier in the book:

1. Start a new project, name it **weather report**, and save it in the project4_weatherman folder.
2. In the open project, switch to a workspace that displays the Project panel, such as the Assembly workspace.
3. Import the two files weatherReport.mp4 and hiking.jpg into the Project panel; leave the other files alone for now.

Importing layered Photoshop documents

★ ACA Objective 4.2

Now you'll import the two Photoshop files, and you'll see that they import slightly differently than the other files.

1. Import the weatherMap.psd file.

    The Import Layered File dialog box appears. It lists the layers inside the document along with options for how to import them. You can import the layers as a single image or as individual images.

2. In the dialog box, click the Import As menu and choose Individual Layers (**Figure 4.1**); then click OK.

    Check boxes let you choose which layers to import, but in this case you want all the layers, so leave them all them selected.

3   Click OK. Notice that the weatherMap.psd file was imported as a bin, with separate images inside the bin that were derived from each layer in the Photoshop file (**Figure 4.2**).

Layered Photoshop files are useful for video graphics because it's possible to animate each layer independently in Adobe Premiere Pro CC. You'll soon see how this works.

4   Import the BBLogo.psd file.

5   Click the Import As menu, choose Merge All Layers to import the file as a single image, and then click OK.

The logo was imported as a single image because there is no need to work with its layers separately in Premiere Pro.

Figure 4.1 Importing the layered PSD file as individual layers

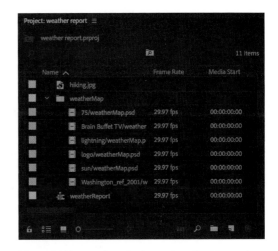

Figure 4.2 Images from the layered weatherMap.psd file imported into a bin inside the Project panel, shown in List view

White balancing a clip using a gray target

Create a new sequence based on the weatherReport.mp4 clip, using any of the techniques you've learned.

The weatherReport.mp4 clip begins with someone holding up a target with three shades of neutral gray on it. Why does this clip start this way? It's a **green screen** clip, which features a subject against a solid green backdrop that you will soon replace with a different background. The process of removing a solid-colored background from a shot is called **chroma key compositing**, or "keying out" the background. *Chroma* refers to the color that's keyed out.

In a natural scene, the camera can usually find some neutral areas to use as a reference for white balancing the video. In a green screen clip, there are neither neutral colors nor natural colors. If the camera is set to automatic white balance, it has no reliable way of deciding what the proper white balance should be. The gray target provides the necessary neutral reference in the clip when you want to use the White Balance eyedropper in the color correction tools in Premiere Pro, such as the Fast Color Corrector you used in Chapter 3 or the Three-Way Color Corrector demonstrated in the tutorial video for this chapter.

Why is such an unnaturally green color chosen as the key color? That color is easy for Premiere Pro to isolate and remove cleanly, without removing anything you want to keep visible. If you use a backdrop with a color that appears in nature, such as an earth tone, a background replacement feature might accidentally remove natural areas you want to keep visible, such as a person's face or clothing.

The target in the video has three strips representing highlights, midtones, and shadows. If you applied a color correction effect that has just one eyedropper, it may be best to click middle gray or white. If you applied a color correction effect with eyedroppers for different tonal ranges, such as the Fast Color Corrector (**Figure 4.3**), you'll want to do the following:

- Click the White Level eyedropper on the white stripe in the handheld target in the Program Monitor.
- Click the Gray Level eyedropper on the middle gray stripe.
- Click the Black Level eyedropper on the black stripe.

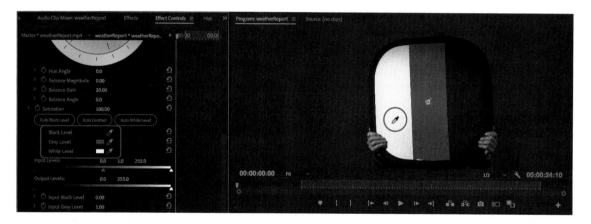

Figure 4.3 The Fast Color Corrector is an effect that uses three eyedroppers for more precise white balancing by sampling highlights, midtones, and shadows.

Preparing to shoot green screen clips

▶ **Video 4.3** *Video Lighting*

A background replacement feature works best when the area to be replaced is cleanly defined so that it's easy to isolate. The background you're replacing must have consistent color and consistent lighting.

Follow these guidelines for successfully shooting a scene on a green screen background:

- Light the background evenly. If you have "hot spots," you may need more lights to cover more of the background area, or you may need to add diffusers to the lights.
- Make sure the green screen is clean and not wrinkled. It should be solid and not contain any kind of a pattern or gradient. This is easy to achieve because you can buy rolls of green screen background or use green screen paint.
- Position the talent several feet away from the background. This will help prevent shadows from falling on the green screen and prevent green reflections (spill) on the talent. Doing so will also make it more likely that the background will be out of focus so that stains or wrinkles on the green screen will be less visible.
- Use a standard key light and fill light to make sure the talent is well lit. (A key light is about lighting the subject, not about chroma keying.)
- Add hair lights so that the rim light effect helps separate the outline of the talent from the background.
- Dress the talent in colors that are not similar to the green screen so that keying software can easily distinguish the background color that needs to be removed.

Compositing a Green Screen Clip with a New Background

★ *ACA Objective 2.3*

★ *ACA Objective 4.6*

▶ **Video 4.4** *Key Weatherman over Weathermap*

Now you're ready to remove the green screen, revealing the weather map underneath.

Drawing a garbage matte

The first phase of green screen compositing is to draw an **opacity mask**, which is traditionally called a **garbage matte**. Although it is possible to simply have Premiere Pro remove the background based on the green color, using an opacity mask quickly masks off the areas that never need to be shown at any point in the clip, reducing the amount of potential variation in the green screen color and making background removal easier and faster.

1. In the Timeline panel, drag the green screen clip, weatherReport.mp4, to a higher track. For example, put a green screen clip on V2 so that you can put the new background under it on track V1.

2. Drag the weather map graphic, Washington_ref_2001/weatherMap.psd, to track V1 at the beginning of the clip.

3. With the Rate Stretch tool, stretch Washington_ ref_2001/weatherMap.psd to match the duration of weatherReport.mp4 (**Figure 4.4**). You need to use the Rate Stretch tool because Premiere Pro sees the PSD file as more of a video clip than a still image.

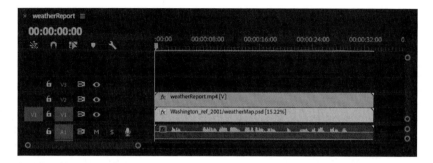

Figure 4.4 The weatherMap graphic set up with the correct position and duration in the Timeline panel

4. Make sure the weatherReport.mp4 clip is selected in the Timeline panel.

5. Scrub through or play back the sequence and note how far out weather reporter Joe's hands extend during the presentation.

6. In the Effect Controls panel, expand the Opacity setting.

7. Select the Free Draw Bezier tool ( ) (**Figure 4.5**).

Figure 4.5 The Free Draw Bezier tool selected in the Effect Controls panel and positioned over the Program Monitor

8. In the Program Monitor, click the Free Draw Bezier tool around Joe to draw a rough mask that stays outside the furthest reach of Joe's hands during the presentation (**Figure 4.6**).

   TIP

   If you've used the Pen tool in other Adobe applications such as Adobe Illustrator or Adobe Photoshop, you already know how to use the Free Draw Bezier tool in Premiere Pro.

Figure 4.6 Drawing the path of an opacity mask

   The mask doesn't have to follow Joe's outline tightly or precisely; leave a bit of margin between Joe and the mask path. Click only in the green areas; don't click any points inside Joe, and don't let any path segments cross over Joe.

9. When you're ready to close the path, click the tip of the Free Draw Bezier tool on the first point you drew.

   The path automatically closes. The area outside the mask becomes transparent (**Figure 4.7**).

Figure 4.7 A completed opacity mask path

10. Play back the sequence and see if any part of Joe's body crosses over the mask at any time.

11. If you need to move a path point or make other adjustments to the path, use the Selection tool to reposition any points on the mask path. If you need to move a point outside the frame, zooming out will let you see outside the frame.

12. If you want to convert any straight segments to a curved segment, Alt-drag (Windows) or Option-drag (macOS) a point to extend Bezier handles. These handles curve the segments extending from a point (**Figure 4.8**).

Figure 4.8 Creating curved segments by extending Bezier handles

Chapter 4 Compositing with Green Screen Effects 209

KEYING OUT THE GREEN BACKGROUND

With the garbage matte in place, now you can have Premiere Pro concentrate on isolating and removing the green screen color that remains.

1. In the Effect Controls panel, find the Ultra Key video effect (remember to use the search feature in the Effects panel). Drag Ultra Key from the Effect Controls panel and drop it on the weatherReport.mp4 clip in the Timeline panel.

2. In the Effect Controls panel, scroll down to the Ultra Key settings and select the Key Color eyedropper (**Figure 4.9**).

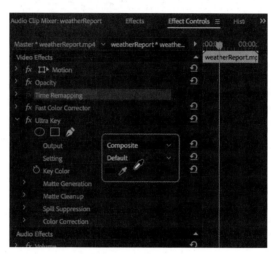

Figure 4.9 The Key Color eyedropper selected in the Ultra Key settings; the pointer is now an eyedropper.

3. Click the Key Color eyedropper on the green screen in the Program Monitor.

   This samples the green color that Ultra Key should remove, and what was the green screen color should now be transparent (**Figure 4.10**).

Figure 4.10 Before and after clicking the green screen color with the Key Color eyedropper

You may get better results if you click the Key Color eyedropper in a darker area of the green screen.

4. In the Effect Controls panel, go to the Ultra Key settings, click the Output menu, and choose Alpha Channel. This displays the mask created by Ultra Key so you can see whether it is clean enough (**Figure 4.11**).

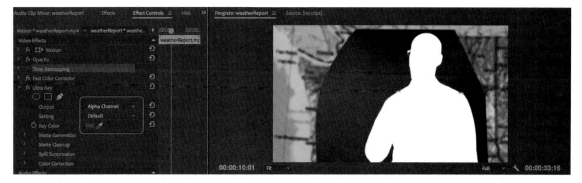

Figure 4.11 After choosing the Alpha Channel setting

There is a saying that can help you remember how to read an alpha channel: "White reveals, black conceals." White mask areas allow the clip to display because they represent opaque areas, whereas black areas are part of the mask that makes those areas of the clip transparent. Gray areas are partially transparent, so dark gray areas are mostly transparent but still let through some of those clip areas.

5. Scrub through the sequence to see if the mask is clean for the entire sequence.

6. If the mask is not clean (not fully black) in some areas, in the Effect Controls panel go to the Ultra Key settings, click the Settings menu, and choose a different option to see which one works best. Each option is a preset for the advanced settings below the Key Color option (**Figure 4.12**). If the advanced settings are expanded, you can see how they change when you choose a Settings preset.

A common reason for mask variations is uneven lighting on the green screen.

> **TIP**
> You can address green spill using the Matte Cleanup options in the Ultra Key effect, especially the Choke and Soften settings.

Figure 4.12 Settings presets change the values in the advanced settings.

TIP

It might be worth playing back the sequence with the lower track hidden. Then you can preview the mask against a black background.

If you're feeling adventurous or already have a technical familiarity with keying, you can expand the Matte Generation, Matte Cleanup, Spill Suppression, and Color Correction settings and try adjusting them. If it feels like you have to work too hard to produce a clean mask, the fastest fix may be to sample a different Key Color by repeating steps 3–5.

7. Scrub through the sequence to see if the mask is now clean for the entire sequence. If it isn't, try step 6 again.

8. In the Effect Controls panel, go to the Ultra Key settings, click the Output menu, and choose Composite.

   This displays the composite result of the two tracks plus the mask applied to the upper track.

9. Play back the sequence to make sure it looks right. Watch out for irregularities in the keyed-out area, and keep an eye out for green spill on the subject.

★ ACA Objective 2.4
★ ACA Objective 4.6

▶ Video 4.5
Add Graphics

TIP

When you want to add more than one video or audio track, choose Add Tracks from the same context menu or choose Sequence > Add Tracks.

Adding and Animating More Graphics

With the weather presentation composited over the weather map, it's time to add some more graphics to help round out the weather report.

Adding a track

You'll soon add graphics on another track. If your sequence doesn't have an empty track above the sequence, add one.

Right-click (Windows) or Control-click (macOS) the Timeline panel just to the left of where the highest video clip starts, and choose Add Track (**Figure 4.13**). Premiere Pro adds a new video track above the track where you clicked.

Figure 4.13 Adding a track

Adding an animated logo

You need to add the Brain Buffet logo so that it enters the frame at the bottom-left corner while rotating and stops at the bottom-right corner of the frame.

1. In the weatherMap bin in the Project panel, drag BBLogo.psd to the beginning of the sequence, onto an empty track above the others.
2. Use the Rate Stretch tool ( ) to display the logo during the entire duration of the sequence.
3. Select the logo and resize and reposition it in the bottom-right corner of the frame, in one of the following ways:
   - In the Program Monitor, double-click the logo, drag to reposition it, and drag its handles to resize it. Be careful not to accidentally drag the anchor point ( ).
   - In the Effect Controls panel, with the Motion settings expanded, make sure Uniform Scale is selected and scrub the Position and Scale values.
4. In the Effect Controls panel, adjust the Opacity value to around 70% to make the logo semitransparent (**Figure 4.14**).

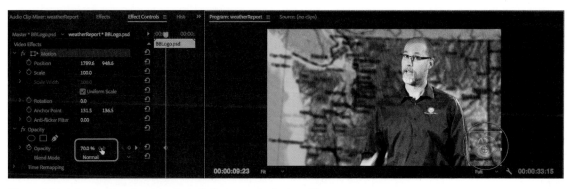

Figure 4.14 Adjusting Opacity of the Brain Buffet logo at its final position

5. Move the playhead to the time when the logo should stop at the bottom-right corner, when Joe finishes saying "Welcome to Brain Buffet TV!"
6. In the Effect Controls panel, enable the Toggle Animation button for the Position and Rotation options so that it adds Position and Rotation keyframes at the current time (**Figure 4.15**).

Figure 4.15 Position and Rotation keyframes added at the playhead

NOTE

Because you rotated counterclockwise, it's normal for the Rotation angle to be a negative value.

TIP

As you add tracks to a sequence, you might want to drag the horizontal dividers between tracks and the audio/video track sections so that you can see what you want to work on.

Figure 4.16 Scrubbing to set the Rotation angle

7 In the Timeline panel, move the playhead to where Joe starts saying "Welcome to Brain Buffet TV!"

8 In the Program panel, if the logo isn't selected, double-click it, and then Shift-drag it to the left until it is slightly off screen. This is its starting point.

9 With the logo still selected, in the Effect Controls panel scrub the Rotation setting to the left to "wind up" the rotation in a counterclockwise direction to its starting point (**Figure 4.16**). Stop when the rotation angle is around –360 degrees.

Because Toggle Animation is on for Position and Rotation, a keyframe is added at the playhead.

TIP

When editing a keyframe, make sure the playhead snaps to it before editing the keyframe value. You know you're on it when Add Keyframe is blue.

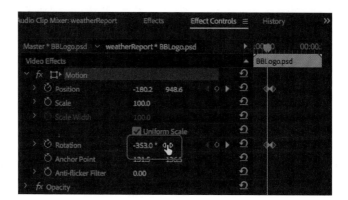

10. To make the logo slow to a stop instead of suddenly stopping, right-click (Windows) or Control-click (macOS) the second Position keyframe, and choose Temporal Interpolation > Ease In from the context menu that appears (**Figure 4.17**).

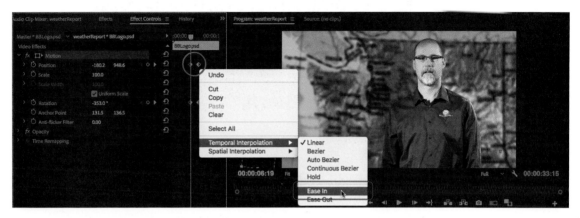

Figure 4.17 Applying the Ease In command to the end of the motion

11. Play back the sequence and evaluate both the movement and rotation. Make any additional adjustments that are needed.

SPECIFYING ROTATION ANGLES

When entering an angle for Rotation, 360 degrees is a complete rotation in one direction, whereas –360 degrees is one complete rotation in the opposite direction. You can go beyond 360 degrees if you want to specify multiple rotations; for example, 1x20 means one complete clockwise rotation plus 20 degrees, and –2x-231 means two complete counterclockwise rotations plus another 231 degrees counterclockwise.

Adding weather graphics to the map

With the presenter now composited over the weather map, it's time to add the weather graphics that appear over the map: a lightning icon and a sun icon. They don't need to move, so to help keep them organized you'll first create a sequence that contains them both.

▶ **Video 4.6** Create Picture-in-Picture

1. In the weatherMap bin in the Project panel, create a new sequence based on the sun/weatherMap.psd file.
2. Rename the new sequence **sun and 75**.
3. From the weatherMap bin, drag the 75/weatherMap.psd file to the Timeline panel. In the Project Monitor, the 75 should appear just below the sun (**Figure 4.18**).

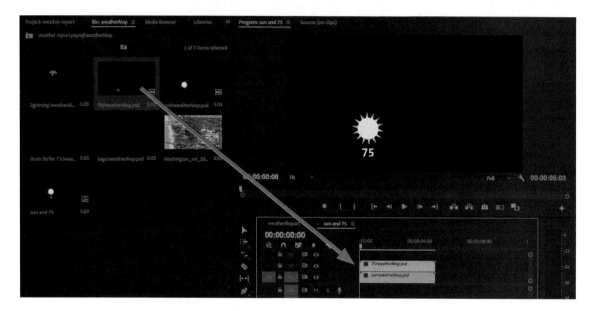

Figure 4.18 Adding the temperature to the sun

4. Click the weatherReport tab in the Timeline panel to make it active.
5. Play the weatherReport sequence so that you can identify where Joe says "sunshine and 75 degrees." Position the playhead to the time where he starts saying that phrase, because that's where you're about to add the sequence you just created.
6. Drag the "sun and 75" sequence from the weatherMap bin and drop it in the Timeline panel, in the empty space just above the top track so that it begins at the playhead (**Figure 4.19**).

Figure 4.19 The "sun and 75" sequence nested within the weatherReport sequence

Dropping an item into the empty space above tracks automatically adds a track for the item you drop, so you don't have to use the Add Track command in advance.

7. Add cross-dissolve video transitions to the start and end of the "sun and 75" clip to fade it in and out.

8. Play the weatherReport sequence so that you can identify where Joe says "thunder and lightning." Position the playhead to the time where he starts saying that phrase, because that's where you're about to add the lightning graphic.

9. Drag the lightning/weatherMap.psd graphic from the weatherMap bin and drop it in the Timeline panel to the top track so that it begins at the playhead.

10. With the lightning/weatherMap.psd graphic selected in the Timeline panel, in the Effect Controls panel scrub the two Motion values (X and Y) to position the lightning graphic over Joe's hand (**Figure 4.20**). This is another way to position a clip.

Chapter 4 Compositing with Green Screen Effects 217

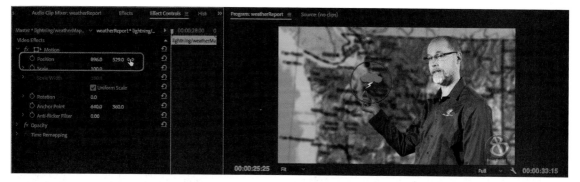

Figure 4.20 Adjusting the Position keyframe values for the lightning graphic

11 Animate the lightning graphic so that it slides down into the frame from the top to follow how Joe's hand comes down when he mentions thunder and lightning.

You did this earlier in this chapter with the Brain Buffet logo, and you can use a similar technique here by using the Effect Controls panel to set Position keyframes at the start and stop of the animation. But this time the primary movement is along the y axis (the second Position option).

12 Play the weatherReport sequence so that you can identify where Joe says "camping." Position the playhead to the time where he starts saying that phrase, because that's where you're about to add a hiking photo.

13 From the Project panel, drag the hiking.jpg file to the top track.

14 Resize and reposition hiking.jpg so that it appears to sit on Joe's hand.

15 In the Effect Controls window, find the Drop Shadow video effect (it's in the Video Effects > Perspective group). Drag Drop Shadow from the Effect Controls panel and drop it on the hiking.jpg clip in the Timeline panel.

TIP

This sequence contains more tracks and effects than you've worked with so far, which is more work for your computer's processor. If playback isn't smooth, click the Select Playback Resolution menu and choose a resolution lower than Full, such as 1/2.

TIP

If scrubbing requires dragging a long way to reach the values you want, press Shift while scrubbing.

16 In the Effect Controls panel, find the Drop Shadow settings until you like how it looks (**Figure 4.21**).

Figure 4.21 Drop Shadow settings in the Effect Controls panel

17 Play back the weatherReport sequence and clean up any loose ends that you find.

If you didn't edit out the gray target at the beginning, move the playhead to just before Joe says "Welcome to Brain Buffet TV!" Set a sequence In point, and set a sequence Out point after Joe stops talking at the end.

Exporting Final Video and Audio

When you're satisfied with how the sequence looks, export the final video using the H.264 YouTube 720 HD preset for easy online delivery and playback to the audience at the retirement home. As you did in Chapter 2, export the sequence to the Adobe Media Encoder queue for final rendering. But before you click the Render button, remember that one of the deliverables is an MP3 audio file that can be used to create a transcript for the hearing-impaired. You can easily create the audio file from the same sequence in Adobe Media Encoder so you don't have to export twice from Premiere Pro.

▶ **Video 4.7** Export with Adobe Media Encoder

To set up creation of the MP3 audio file:

1 In Adobe Media Encoder, select the weatherReport sequence you exported from Premiere Pro, and click the Duplicate button (**Figure 4.22**).

Figure 4.22 Duplicating the weatherReport sequence

2 In the Preset Browser panel, expand the System Presets list and then expand the Audio Only list.

3 Drag the MP3 128Kps preset from the Preset Browser, and drop it on the duplicated sequence (**Figure 4.23**). The Format and Preset for the duplicate sequence change to indicate the new settings.

Figure 4.23 Dropping a preset onto a sequence

When you click the green Start Queue button in the Queue panel, Media Encoder will process the items in the queue, producing H.264 video and MP3 audio versions of the sequence for you.

> **NOTE**
>
> A Media Encoder queue item doesn't update if you edit the sequence it came from. If you change the content of a sequence and want to render the revised version, you must export it from Premiere Pro again.

Challenge: Create Your Own Composited Video

Now it's time for you to come up with your own special effects video.

As you plan your project, remember Joe Dockery's Keys to Success (from the video):

- Keep it short, around 30–60 seconds long.
- Determine the background media. It can be a different part of your city, a picture or video of an exotic location, or even another planet.
- Determine the background that you'll remove. It can be a large sheet of paper or a painted wall, as long as it's a distinct color that won't be confused with any colors in the content that you want to keep visible.
- Shoot good-quality video and record good-quality audio to minimize the amount of work you have to do in postproduction.
- Frame actors tightly, such as from the waist up, to simplify keying.
- Plan the timing of the talent's lines and movements to coordinate them with other elements you want to composite into the scene.
- Set the white balance of the camera with the gray target.
- Follow the other guidelines for lighting and shooting green screen clips earlier in this chapter.

▶ **Video 4.8** *Special FX Challenge*

Conclusion

In this chapter you've gotten a taste of how Hollywood and television can make any idea look real by conceiving sequences as visual composites of live action video clips, backgrounds, and digital graphics. Let your own imagination run free!

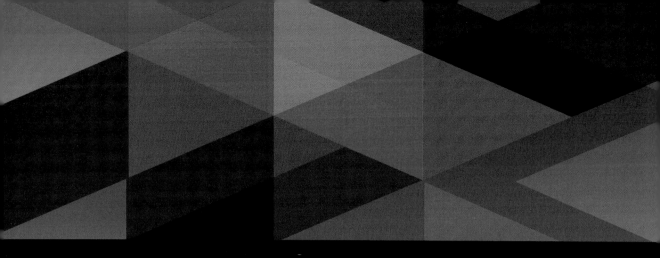

CHAPTER OBJECTIVES

Chapter Learning Objectives

- Review file management and project organization.
- Organize files.
- Open and save a project.
- Create a sequence.
- Set preferences for importing multiple still images.
- Use markers.
- Set up Automate to Sequence to accelerate sequence creation.
- Set up variations of a job in Adobe Media Encoder.

Chapter ACA Objectives

For full descriptions of the objectives, see the table on pages 279–283.

DOMAIN 1.0
WORKING IN THE VIDEO INDUSTRY
1.1, 1.4

DOMAIN 2.0
PROJECT SETUP AND INTERFACE
2.1, 2.3, 2.4

DOMAIN 3.0
ORGANIZATION OF VIDEO PROJECTS
3.1

DOMAIN 4.0
CREATE AND MODIFY VISUAL ELEMENTS
4.5, 4.6

DOMAIN 5.0
PUBLISHING DIGITAL MEDIA
5.1, 5.2

CHAPTER 5

Creating a Video Slide Show

Your latest assignment is something that Brain Buffet is often asked to do: create a short memorial video of a deceased loved one or community member from photographs provided by family and friends. Using tools in Adobe Premiere Pro CC, it doesn't take long to set many photos to music and add motion effects. The result is a moving presentation that tells the story of why and how the deceased should be remembered.

Preproduction

As you've learned in previous chapters, production starts only after the project requirements are clearly understood and agreed on, so let's review those before you begin:

- **Target audience:** Family and friends of the deceased, from 10 to 80 years old.
- **Goal:** Celebrate the life of a family or community member who recently passed away.
- **Deliverable:** The client expects a 1-to-3-minute video of photographs set to music, in two versions: one version optimized for fast online delivery and another version optimized for high-quality playback on a computer at the memorial service.

The available media files are photographs of the departed and background music for the slide show.

★ *ACA Objective 1.1*

★ *ACA Objective 1.4*

▶ **Video 5.1**
Introducing the Memorial Slide Show Project

★ ACA Objective 2.1

★ ACA Objective 2.4

▶ **Video 5.2**
Organize Your Project

Setting Up a Slide Show Project

Because this slide show contains many images, it's a good idea to decide on the default duration for still images so that you need to set it only once. For this project you want each slide to display for 4 seconds, and the place to set that duration as the default is in the Premiere Pro Preferences dialog box.

1. Open the Preferences dialog box and select the Timeline panel.
2. For Still Image Default Duration, choose Seconds and then enter **4** (**Figure 5.1**).
3. Click OK.

Figure 5.1 Setting the Still Image Default Duration value

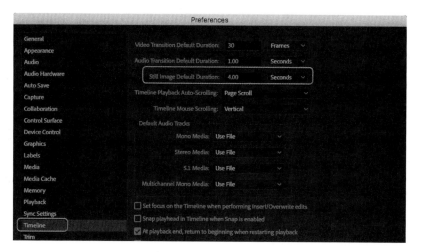

Start the editing stage of production by practicing the project setup techniques you've been using throughout the book:

1. Unzip the project files and store them in a folder you create for this memorial project.
2. Start a new project, and name it **Memorial**.
3. Save the project into the Memorial project folder you created.
4. Switch to a workspace that displays the Project panel, such as the Assembly workspace.

The images are all in a single folder, and that makes them easy to import all at once. Import the Video Clips folder into the Memorial project using your favorite import method. Because you imported a folder instead of one file, the folder is added as a bin within the Project panel.

Import the Memorial.wav file into the Project panel, not inside the Video Clips bin. Memorial.wav is the music for the slide show.

Creating a Sequence from Multiple Files Quickly

★ ACA Objective 2.1
★ ACA Objective 2.3
★ ACA Objective 3.1

There are 50 still images in the Video Clips folder for this project. Dragging each of them individually to the Timeline panel, adjusting their durations, and adding transitions to each clip could take a long time. Fortunately, Premiere Pro has ways to speed up adding multiple items to a timeline and adding transitions to them.

Creating a sequence based on a preset

Up to this point, most of the new sequences you've created have been based on one of the source video clips. You won't be able to set up this project's sequence that way, because there are no source video clips; they're either still images or audio. You'll need to start a new sequence from scratch and specify what sequence settings it should use.

1   Start a new sequence using any of the methods you have learned so far.

2   In the New Sequence dialog box, name the sequence **Memorial Slide Show**, but don't click OK yet.

    You'll choose a preset that's closest to the high-definition frame dimensions and high level of quality that the slide show should have when it's played on a computer, which is the higher-quality deliverable in the preproduction requirements.

3   In the Available Presets list, expand the Digital SLR group; then expand the 1080p group and select DSLR 1080p30 (**Figure 5.2**).

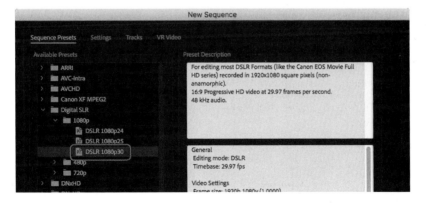

Figure 5.2 Choosing a preset to define the sequence settings

The DSLR 1080p30 preset uses the popular and high-quality standard of 1080p video (1920x1080-pixel frame size, progressive scan) at 30 frames per second.

4 Click OK.

Now you'll add the music file. It will define the duration of the slide show.

5 Add the Memorial.wav music file to audio track A1 in the Timeline panel, beginning at the start of the sequence.

You'll need two copies of the sequence for later, so you'll duplicate one of them now.

6 In the Project panel, select the memorial slide show project and choose Edit > Duplicate.

7 Click the filename of the duplicate and name it **Markers**. You'll work with this sequence later.

Next, you'll try two ways to add media to a sequence that are much faster than adding and editing them one by one. You'll practice these techniques with photos, but keep in mind that they'll work with video too.

Arranging multiple items before adding them to a sequence

▶ *Video 5.3*
Organize Your Photos in the Project Panel

One way to save time is by using the Project panel or bin as a sort of storyboard pad for getting clips in the right order before you add them to a sequence.

1 Position the pointer over the Project panel or opened bin that contains the media you want to arrange, and press the tilde (~) key.

Maximizing the panel isn't required, but it lets you see the most items at once as you arrange them, especially if you're working on a small display.

2 Drag image thumbnails to put them in the order that tells the story best.

3 When you're done arranging, position the pointer over the Project panel or opened bin that contains the media you were arranging, and press the tilde (~) key to restore the panel to its original size.

The next steps will use the order of media you've arranged here.

Adding multiple items to a sequence at a regular interval

Use this technique when you want to add selected media items to the timeline with the same interval between the start of each item. In this case the interval is based on the Still Image Default Duration value you set earlier in Preferences.

▶ *Video 5.4
Automatically Add Photos to the Timeline*

1. Make sure the memorial slide show sequence is active in the Timeline panel and that the playhead is at the beginning of the sequence.
2. Make sure the Video Clips bin is active, and choose Edit > Select All.
3. Click the Automate to Sequence button ( ).
4. In the Automate To Sequence dialog box, set the following options (**Figure 5.3**):
   - Click the Ordering menu and choose Sort Order.
   - Click the Placement menu and choose Sequentially.
   - Make sure the Clip Overlap option is set to 30 Frames. This ensures that clips overlap enough that you can apply a transition.
   - Make sure Apply Default Video Transition is selected.

   TIP
   If you want to change the default transition that's applied between items on the Timeline, do that before you start these steps. In the Effects panel, expand the Video Transitions, right-click (Windows) or Control-click (macOS) the transition you want, and choose Set Selected As Default Transition.

5. Click OK.

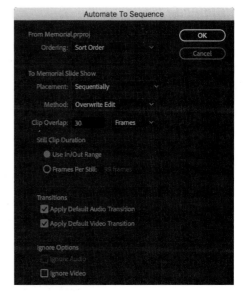

Figure 5.3 Setting up the Automate To Sequence dialog box for sequential placement

The clips are added to the sequence using the Still Image Default Duration value specified in Preferences, using the sort order you used and applying the default transition between each item. That entire sequence took only a few moments to set up.

TIP
The Replace With Clip > From Bin command is also available if you right-click (Windows) or Control-click (macOS) an item in the Timeline panel.

If you want to adjust any of the edits or transitions, you can do that using the editing techniques you've learned in earlier chapters.

You may find that you want to replace an item in the sequence with another item. Here's how you do that:

1 In the Project panel or bin, select the item you want to use.
2 In the Timeline panel, select the unwanted item in the sequence.
3 Choose Clip > Replace With Clip > From Bin.

Adding multiple items to a sequence at markers

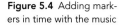
Video 5.5 *Use Markers to Place Photos*

If you want more control over where Automate To Sequence adds items to a sequence, you can use markers. For example, if you want to pace the slide show images to music, you can add sequence markers at key moments in the music.

1 Open the Markers sequence you created earlier as a duplicate of the Memorial Slide Show sequence.
2 Move the playhead to the beginning of the Timeline panel, and press the M key to add a marker.
3 Position a finger over the M key; be ready to press the M key during playback.
4 Play back the sequence.
5 When you hear a point in the music where a new image should appear in the slide show, press the M key. A marker appears on the timeline.
6 Continue dropping markers until the music ends (**Figure 5.4**). If you think you might have missed a time when you should have dropped a marker, feel free to replay any part of the sequence and add missing markers.

Although there is a button you can click to add a marker, it's often easier to use the M key shortcut instead.

Figure 5.4 Adding markers in time with the music

7. If a marker is not exactly at the correct frame, simply drag it left or right until it's at the correct time.

    Now you're ready to add the photos to the sequence.

8. Make sure the markers sequence is active in the Timeline panel and that the playhead is at the beginning of the sequence.

9. Make sure the Video Clips bin is active, and choose Edit > Select All.

10. Click the Automate To Sequence button.

11. In the Automate To Sequence dialog box, set the following options (**Figure 5.5**):

    - Click the Ordering menu and choose Sort Order.
    - Click the Placement menu and choose At Unnumbered Markers.

    Many other options aren't available because they apply only when Placement is set to Sequentially.

12. Click OK.

Figure 5.5 Setting options in the Automate To Sequence dialog box

The clips are added to the sequence using the Still Image Default Duration value specified in Preferences and in the sort order you used. Each clip is added to the sequence at the next available unnumbered marker.

You may see gaps between items. This can happen for photos when the Still Image Default Duration value specified in Preferences is shorter than the time between markers. You can easily fill those gaps by dragging one end of a photo with the Selection tool to extend is duration until it snaps to the adjacent clip.

Adding the default transition between items

If you want to add the default transition between items, you can still do that quickly, although it's a manual process:

1. Navigate to the next edit by pressing the Down Arrow key.
2. Press Ctrl+D (Windows) or Command+D (macOS).

    That's the keyboard shortcut for the Sequence > Apply Video Transition command. Although it's usually easier to use menu commands when you're learning, in this case it's going to be a lot easier and faster to keep your hands on the keyboard pressing the Up Arrow key and the transition shortcut than it would be to keep going up to the menu bar every time you want to add a transition.

> **TIP**
> You can select all of the images you added to a track and choose Sequence > Apply Default Transition. This adds the default transition between all selected items. Be sure to review each edit point because a transition may not be applied where a gap exists.

3. Repeat steps 1 and 2 to add the default transition (**Figure 5.6**) at any other edit where it's needed. (If you want to go to the previous edit, press the Up Arrow key.)

Figure 5.6 Adding the default transition at each edit

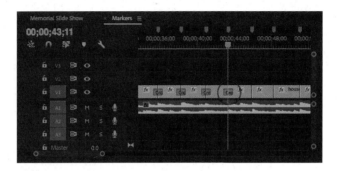

TIP

You can add a text note to a marker by double-clicking it.

★ ACA Objective 4.5

★ ACA Objective 4.6

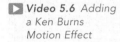 **Video 5.6** Adding a Ken Burns Motion Effect

Adding a Ken Burns Motion Effect

The Ken Burns effect refers to panning and zooming the camera over a still image so that it produces a more dynamic effect than a motionless image on the screen. The effect is named after the documentary filmmaker Ken Burns. He did not invent the technique, but it was noted as an effect he frequently used with old photos and artifacts in his historical documentaries.

Now that you know that the Ken Burns effect is about panning and zooming, hopefully you have already guessed how you can easily achieve that effect in Premiere Pro. You can pan and zoom by setting keyframes for the Motion options in the Effect Controls panel when a clip or still image is selected in the Timeline panel—a technique you've practiced in earlier chapters. You control panning with Position keyframes, and you control zooming with Scale keyframes (**Figure 5.7**).

Choose a few images in the Markers sequence to practice Ken Burns–style panning and zooming. As you do this, the following guidelines will help:

- For panning across a photo, you need only two Position keyframes if you want to move the camera in a straight line. You can add more keyframes if you want to move the camera in different directions.
- For zooming a photo, you usually use only two Scale keyframes.
- Avoid sudden changes in scale or position unless there's a reason.

Figure 5.7 The frame starts showing the package being handed to the veteran and ends after panning his head closer to the center and scaling up the frame to zoom in on his reaction. The keyframes used are shown in the Effect Controls window.

- When you right-click (Windows) or Control-click (macOS) a keyframe, the Temporal Interpolation commands can help smooth changes in speed. The Ease In and Ease Out commands smooth stopping and starting, respectively. The Spatial Interpolation commands can help smooth changes in position.

- If you plan to zoom into an image by a high magnification, use a frame size larger than the video frame size. For example, if you want to zoom in by 2x for a 1920x1080-pixel video frame, you should use a version of that image with twice the width and height of the video frame (3840x2160 pixels) if available so that the image maintains maximum detail when zoomed in.

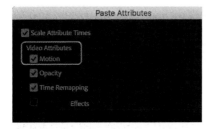

Figure 5.8 Select the Motion check box under Video Attributes.

If you plan to use similar motion effects on multiple photos, you can copy and paste the effect to use as the starting point for the next photo:

1 Apply the effect to the first photo.
2 With the photo selected, choose Edit > Copy.
3 Select another clip and choose Edit > Paste Attributes.
4 In the Paste Attributes dialog box, select the Motion option under Video Attributes (**Figure 5.8**), and click OK.
5 Customize the effect for that photo.

Copying and pasting attributes and effects works for video, still images, and audio.

★ ACA Objective 5.1
★ ACA Objective 5.2

 **Video 5.7** *Export Your Slideshow*

Exporting Multiple Versions with Adobe Media Encoder

The requirements for this project specify two versions: one version optimized for fast online delivery and another version optimized for high-quality playback on a computer at the memorial service. You have two different versions of the sequence to provide in both formats. But you don't have to export from Premiere Pro four times, because you can use Adobe Media Encoder CC, a tool you first met in Chapter 1.

You'll set up the initial export in Premiere Pro. This first export will use the same settings as the sequence, which will produce a high-quality version.

1 Open the memorial slide show sequence.
2 Choose File > Export > Media.
3 Click the Format menu and choose H.264.
4 Click the Preset menu and choose Match Source - High Bitrate.
5 In the Video tab, click the Bitrate Encoding menu and choose VBR, 2 pass (**Figure 5.9**).

Two-pass encoding can produce higher quality for the same file size compared to one-pass encoding, so it's being used here for the high-quality slide show. However, two-pass encoding takes more time to process.

Figure 5.9 Customizing export settings

If you can't see the Bitrate Encoding option, make sure you clicked the Video tab. It's near the bottom of the Video tab, so enlarge the Export Settings window or scroll down within the Video tab.

6 Click Output Name and make sure this first export will be saved to the location and filename that you want. Add **HQ** to the end of the filename to signify that this will be the high-quality version.

Keep in mind that you'll be rendering multiple versions of this sequence. Have a plan for where to store them all (should they be in a new folder of their own?) and what to name each of them so that you can tell them apart.

7 Click the Queue button to send the sequence to Media Encoder.

8 Open the Markers sequence, and repeat steps 1–7.

Now you'll switch to Adobe Media Encoder and set up the other versions of those two sequences there, because it will be less work than doing two more exports from Premiere Pro.

NOTE
If you change any settings after choosing a preset, the Preset menu will change to Custom because you have altered settings away from how the preset is defined. You can save your customizations as your own preset.

9   Switch to Media Encoder.

10 In the Queue panel, select one of the sequences and then Ctrl-click (Windows) or Command-click (macOS) the other sequence so that they're both selected (**Figure 5.10**).

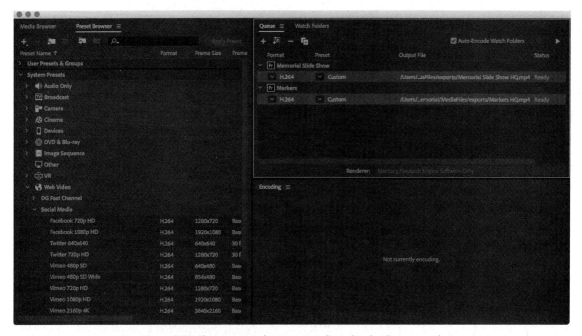

Figure 5.10 The two original sequences selected in the Queue panel

11 Click the Duplicate button (**Figure 5.11**). Both sequences are duplicated; note the filenames to identify the duplicates.

12 Make sure both duplicates are selected (the files with names ending in _1).

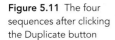

Figure 5.11 The four sequences after clicking the Duplicate button

13 Click the Preset menu (not the text) for either selected sequence and choose YouTube 1080p HD. The applied preset changes for both selected sequences (**Figure 5.12**).

Figure 5.12 After choosing the YouTube 1080p HD preset for both selected sequences

If you just need to change the preset, choose it from the Preset drop-down menu for an item. If you want to customize specific settings, click the Preset text.

14 Deselect the items, and click the Output File text for the first duplicate and edit it so that the filename ends with **LQ** (for Low Quality); then click Save. Repeat for the other duplicate.

15 Click the Play button to start the rendering queue (**Figure 5.13**).

Media Encoder shows you which sequences are processing, the progress of all sequences, and the progress and thumbnail of each sequence that's currently processing.

Figure 5.13 Queue processing in progress

Challenge: Your Own Slide Show

*Video 5.8
Memorial
Challenge*

Now it's time for you to come up with your own slide show. You can create one for a memorial, a birthday, a wedding, a sports team's season, or a special event.

As you plan your slide show, remember Joe Dockery's Keys to Success (from the video):

- Keep it short.
- Allocate plenty of time for collecting photographs and other special objects and for scanning or photographing items that aren't already digital.
- Find appropriate music for the background.
- Add a little movement to the images to maintain visual interest.
- Share it with the world.

Conclusion

A project like a slide show can be effective with little or no video at all, especially because you can add motion by animating the images. Being able to transform still images into a compelling slide show is another tool that you can use to tell stories with Premiere Pro.

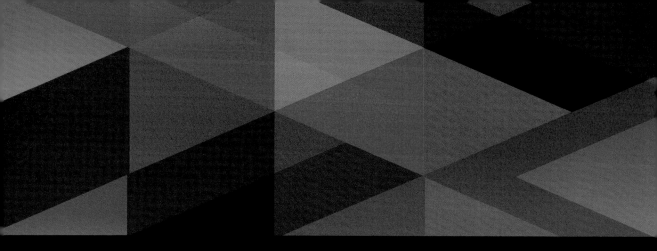

CHAPTER OBJECTIVES

Chapter Learning Objectives

- Understand the phases of production.
- Understand job requirements.
- Understand the roles of a video production team.
- Communicate effectively.
- Use different shots used in cinematography.
- Compose the video frame.
- Understand editing techniques.
- Be aware of intellectual property rights and licensing.

Chapter ACA Objectives

For full descriptions of the objectives, see the table on pages 279–283.

DOMAIN 1.0
WORKING IN THE VIDEO INDUSTRY
1.1, 1.2, 1.3, 1.5

CHAPTER 6

Working in the Video Industry

Although many people capture and edit video by themselves, video and film production has traditionally been a team effort. When you're doing video as a business, working as a team is necessary to achieve the production values that clients expect to get for their money. Producing that level of quality generally involves too many details for one or even two people to manage.

This chapter gives you an overview of the phases of production, takes a closer look at defining job requirements, introduces you to the various roles of a production team, and provides a guide to common shots and editing techniques.

Phases of Production

Whether it's a short public service announcement or a theatrical feature, a professional video production typically moves through the following stages.

★ *ACA Objective 1.2*

Development

The development phase is where you'll review the job requirements discussed earlier. Much of the development phase is about agreement, commitment, and strategy—the high-level stuff that you'll need in order to move to the next level of detail in the preproduction stage that comes next.

- **Agreement.** Reviewing the job requirements with the client makes it possible for both of you to agree on what is to be accomplished by producing the video.
- **Commitment.** Working out the budget and the schedule with the client, and securing funding for the production, ensures that everyone involved will contribute the money and time needed to get the project done. Getting time and money commitments from all involved parties allows the project to move forward to preproduction.
- **Strategy.** The high-level job requirements defined by the client, target audience, purpose, and deliverable start to drive the more detailed decisions that will be made in preproduction.

Preproduction

You can't just rush into production after development. If you simply grab some people and start rolling cameras, too much will be missing. Why are there no lights or costumes? Weren't we supposed to shoot the ending back at the other location we already left? How come we ran out of memory cards? An important purpose of preproduction is to avoid all surprises that might cost more time and money to fix. You achieve this through extremely detailed planning that includes the following steps:

- **Creating storyboards.** The ripple effect of creating storyboards is enormous. The expressions on character sketches remind the director how to guide the actors during each scene. The compositions suggest to the director of photography how he or she will want to frame each shot in camera. The storyboard sequence informs editors how they'll want to put the captured shots together in postproduction. Notes on the storyboards guide production specialists such as the costume designer, art director, and location manager.
- **Writing the script.** The script takes the storyboard and fills in another level of detail. For the actors, the script is the level of detail they need, because it contains the exact dialogue they will speak. Others on the production team use the script as a more specific guide to the details they will need to fill in.
- **Hiring.** Actors are auditioned, and members of the production team are interviewed and hired.
- **Planning production details.** If the script says the characters are watching a sunset on a Hawaiian beach in summer, the location manager knows to find a west-facing beach that can pass for Hawaii, the costume designer knows the

characters need light beach wear for summer, and the director of photography knows to design a lighting setup that's appropriate for the warm colors and long shadows of a sunset scene.

- **Scheduling in detail.** Time needed for a project is directly related to the budget, so accurate scheduling is critical. And scheduling is tied into everything. For example, to shoot a scene on a street corner, the location manager needs to be able to secure a permit on a specific day between specific times. All actors and crew must be available at the time. All equipment must be in place at that time, so logistics such as transportation to the location must be scheduled. All scenes needed on that corner must be shot within that time window, regardless of where they are in the script, so the director and actors must be prepared to shoot those specific scenes. All costumes for that scene need to be finished and ready by that day. Delays and reshoots are expensive largely because of the need to coordinate all those resources again if the original schedule missed a detail.

Preproduction requires a lot of communication between members of the production team, as specialists anticipate potential issues and resolve them through planning. For example, the director of photography might require a certain lighting setup for a specific location, which ends up changing the power system that the electrician must safely design for that location. Other related specialists such as the production designer, costume designer, and location manager will review each other's plans for each scene to understand what they will be expected to provide, or to point out concerns based on schedule or budget.

In addition to a detailed master schedule, production specialists will create schedules and checklists for their own departments to make sure nothing is missed. Once again, everything must be anticipated because one mistake can ruin a scene, and rework impacts the schedule, which impacts the budget.

Production

There is a saying that goes, "Plan the work, then work the plan." That's a succinct description of preproduction, production, postproduction. Preproduction is planning the work, and everything from production forward is working the plan. The production phase is about doing everything exactly as it was laid out and scheduled in preproduction, in the right place, at the right time, with the right people and equipment.

If everything was properly anticipated and scheduled, each shoot should produce video and audio clips that are ready to be edited as the storyboard and script indicated, with no missing pieces. Are all lighting instruments in place? Were all camera cards ingested, logged, and fully backed up? To make sure this happens, all production specialists will follow their own schedules and checklists.

Some productions may involve non-live-action elements such as animation or computer graphics. While the production crew is shooting live action, effects facilities may be working on their synthetic elements so that their clips are also ready for the editor to combine with the live-action elements.

Postproduction

After all necessary assets are captured or created, such as video, audio, and animation, they are brought into the postproduction phase. In postproduction, the elements of the production are edited together and polished.

Using the script as the guide, the editor assembles video and audio into a rough cut. After discussing the rough cut with the director, the editor continues to refine the edit until the story is told smoothly with effective pacing. As the edit becomes more refined, special effects sequences, graphics, and sound effects are added. After the edit is locked down, the editor can add music and titles.

It's during this phase that the editor spends a great deal of time in a video editing application such as Premiere Pro CC, using techniques similar to those you have used in this book.

Distribution

When postproduction is declared complete, the resulting video can be distributed in the form of the deliverables specified in the development phase, whether it's destined to be viewed on an online streaming service, in movie theaters, on television networks, or on an internal corporate website.

For videos that are expected to generate revenue, marketing is a key activity that makes potential paying viewers aware of the production and motivates them to watch it. Marketing and advertising may start before the distribution phase.

Reviewing Job Requirements

★ ACA Objective 1.1

For the sample projects in this book, your preproduction phase involved looking over the requirements for the job. That's because it's critical to be clear about job requirements before you begin; they'll drive all the decisions that you and your team make. This part of the project can be fairly informal on smaller projects but can be huge on large projects. Here's a list of critical questions to answer:

- **Purpose.** Why are you doing this project? What result would you consider a success?
- **Target.** Who needs this message or product? Who is your typical customer?
- **Budget and other limits.** What are the limits for the project? How much budget and time have been allocated?
- **Preferences.** Aside from the results we've already discussed, are there any other results you'd like or expect from this project?
- **Platform.** Is the project targeting the web? A mobile device? A kiosk? DVD? Broadcast TV? What are the specifications of the particular device?

These examples are intended to show how quickly you can determine a client's expectations. The answers to these questions define the size of the job and how you'll best be able to work with the client.

Sketches and written notes from this initial step will help. Gather as much information as you can to make the rest of the project go smoothly. The more you find out now, the less you'll have to redesign later because the client hates the color, the layout, or the general direction you took the project. Invest the time now, or pay it back with interest later. With a clear idea of what the problem is, you'll get the information you need to solve it in the next step.

The range of potential requirements is far larger than the samples covered in this book, so let's look at some of the possibilities.

Client

The client examples in this book included a construction company, a school district, and a retirement home. Potential clients for your video productions might include the following:

- Local businesses
- Sports teams

- Technology startups
- Real estate agencies
- Travel agencies
- News organizations
- Arts organizations
- Restaurants
- Event organizers or promoters
- Schools, universities, and other education organizations
- Governments
- Churches and other religious centers

If you have a special interest, such as food, weddings, animals, aviation, or aerial (drone) videography, you may decide to concentrate on that as a niche. That way, you can develop specialized expertise so that you can provide the best possible video coverage of those areas. Specialization can help differentiate you in the market.

Target audience

Keep in mind that the target audience is often different than the client. For example, the people who hired you might be a technology startup company developing a smartphone app for busy parents. Therefore, the target audience would be the busy parents, not computer programmers.

In some cases, your client may not be clear about the target audience. You should prepare for this possibility with questions that help paint a mental picture of who the target audience is. A good way to do this is to identify demographic attributes of your target audience. For example:

- **Income.** Determine if you want to focus on quality, exclusivity, or price.
- **Education.** Establish the vocabulary and complexity of the design.
- **Age.** Dictate the general appeal, attitude, and vocabulary.
- **Hobbies.** Help in choosing images, insider vocabulary, and attitudes.
- **Concerns, cares, and passions.** Identify core beliefs, trigger points, and so on.

Identifying the target audience demographically helps you and your scriptwriter to determine the imagery, language, look, and atmosphere of your video that would make it more likely for the target audience to relate to your video, decide it's consistent with their interests, and become motivated to pay attention.

Purpose

The reason you get hired is because a client has decided that a specific message needs to be delivered through video. The purpose of the video should be clear enough that you can make the decisions to move the project forward, such as hiring an appropriate scriptwriter for the target audience.

If you meet with the client and their stated purpose for a video is vague, be prepared to ask questions that clarify the purpose. For example, maybe a client wants to make a web ad because an online survey revealed that a low percentage of the target audience was aware of that client's business. You might help the client decide that the purpose of the video is to increase brand awareness among television viewers by 15 percent in the next survey. Again, being more specific will help you make better decisions so that you can plan your production more precisely. Doing so will increase your chances of success.

Deliverable

In the past, a video might have involved one deliverable, such as a videotape. Today, projects have a wider range of potential deliverables:

- Television
- Streaming or downloadable video for the web and mobile devices
- Digital Cinema Package (DCP), a collection of data files in a standard configuration for screening at a movie theater
- Optical media such as DVD or Blu-ray Disc

In addition, the range of technical requirements is broader than ever. Although the majority of your projects may be 2K or 4K video in 2D, some clients may need HDR video, 3D video, or 360-degree video (often called 360 video). Some of these formats have special requirements for capture and editing, making it even more important to communicate clearly with your client.

Many jobs may require only stereo or even monophonic audio, but some may demand a 5.1-channel surround sound mix in one of several surround formats. For example, you may be asked to provide a video for an exhibition installation that will use a specific surround sound system.

In some cases, you may need to partner with a facility that can help process your video and audio to specific delivery.

★ ACA Objective 1.2

Roles of a Video Production Team

A successful video production results from crew members playing many roles in developing the intentions of the director and the script into the visual details that are captured on camera. Not every production team has all of these roles, but most large productions do. And though there are additional roles not listed here (you know how long television and movie credits can be), these are many of the common roles:

- **Producer.** The producer is typically the primary contact with the world outside the production, such as a client, a studio, or the financial backers of a production. This individual may have developed the idea for a production and secured the financing for it. He or she will assemble and manage the production team. An executive producer is a level above the producer and is typically not involved in day-to-day production, but may have provided significant financial backing.

- **Director.** The role of director demands a clear vision and an ability to manage a team. The way the director interprets a script strongly influences all aspects of a production, such as the overall style and mood of a production, because the director's interpretation drives the execution of the acting, cinematography, production design, costumes, and so on.

  A director may be supported by a first assistant director, who helps monitor schedules, checklists, and other production details.

- **Screenwriter.** The screenwriter creates a screenplay, which is a story written in a format that can be directly translated to action items for a production team. The dialogue in the script becomes the lines that the actors speak, and the script elements around the dialogue provide the basis and context for various production tasks. For example, when a scene heading says INT. LONDON PUB, the director knows to find an English pub to shoot in or arrange the construction of an appropriate pub set. Dialogue must be written in a way that allows actors to realize a character through both speech and action.

- **Director of photography.** Also called a DP, this role executes the director's vision of the script through the use of cameras, lenses, composition, and lighting. Naturally, the DP is the primary driver of the visual look of the film.

  A director of photography may be assisted by camera operators, as well as by a digital imaging technician (DIT), who specializes in the technical details of keeping the cameras and their accessories running at all times. For example, the DIT will make sure that all batteries are kept charged, that empty data cards are always available, and that used data cards are ingested, logged, and backed up.

- **Production designer.** Although the director of photography creates the look of a film in camera, he or she has to have something to shoot. Production designers also have a huge influence on the look of a film through what they put in front of the camera. For example, if a production is set in a specific time period, the production designer defines the details that make a location or a set appear to be of that time period.
- **Art director.** The art director realizes the intentions of the production designer by obtaining or building the physical settings for the scenes.
- **Costume designer.** The costume designer's job is similar to the production designer's job but applies specifically to the clothes that the actors wear. Ideally, the costumes combine with the sets and the cinematography to convincingly immerse the audience in a specific time and place as indicated by the script.
- **Talent/actor.** The actor executes the script on camera. An actor employs skills such as speech delivery, facial expressions, and movement to convincingly portray a character. For commercial or documentary projects, the role of talent may be to communicate the script's message in an authoritative and credible voice and manner.
- **Editor.** In postproduction, the editor takes all the video and audio clips that were captured in the production phase and assembles them into the finished video. Much of this involves working closely with the director so that the timing of edits is consistent with how the director intended each scene to flow and coordinates with the acting. When scenes are shot with multiple takes or on multiple cameras, choosing the best shot is also typically the result of the editor and director working together.
- **Production assistant.** There can be multiple production assistants taking care of various tasks that senior production team members are too busy to cover— everything from setting up technical equipment and props to running errands and getting meals for the crew.
- **Gaffer.** The gaffer is the lighting director who creates the lighting plan. Gaffers will work with the key grip, who executes the lighting plan by setting up the equipment.
- **Location manager.** When a production needs to shoot outside the studio, the location manager scouts for places that fulfill the requirements of various scenes in the script. To make it possible for the production to shoot at a

location, the location manager will also obtain any necessary permits and permissions and will organize the logistics that will transport production equipment to and from the location.

- **Unit production manager.** This critical role involves managing and monitoring the schedule and budget. The unit production manager should be able to notice as soon as a production starts to fall behind schedule or exceed its budget so that appropriate steps can be taken to keep the production on course.
- **Other roles.** The bigger the production, the more detailed the roles can be. Larger productions may include grips who set up camera and lighting accessories, camera operators, camera assistants, key grips who set up equipment, audio technicians/sound mixers, boom operators, and electricians.

Teams on smaller productions

Obviously, smaller productions may combine roles, with some members performing multiple roles.

When the budget allows, it's better to break out each role. For example, editing is a skill that's distinct from cinematography. While one person may be able to write, direct, shoot, and edit, having different people in these roles often results in a better product, because having more eyes on the content results in alternate opinions that can produce constructive feedback. Productions that can afford larger crews typically have higher production values, because more skilled specialists can attend to the details in every area of a production.

★ ACA Objective 1.2

Communicating Effectively

Whether a production is large or small, the members of a video production team must be precisely coordinated in order for the production to stay on schedule and within budget. This requires constant clear communication, both vertically (between heads and assistants) and horizontally (across departments).

Everyone will have to agree on matters such as the following:

- **Budget.** The budget defines the resources available to everybody, from actors' fees to equipment and locations.
- **Schedule.** Each important date on the schedule dictates the deadlines for all production roles that need to participate.

- **Aesthetic approach, mood, and atmosphere.** All production roles must support the overall vision of the production. For example, the director, production designer, and costume designer might have different personal ideas of what life might be like in the year 2200, but if they are all working together on a science fiction script set in the year 2200, they must all agree on how that script should be realized. They need to communicate effectively so that they are aware of the work being done by others and can coordinate to create a unified look. If the script tells the story of a utopian civilization in the year 2020, the director, production designer, and costume designer might all work to achieve a visual look and acting style that is bright, clean, free, and optimistic. But if the script describes a dystopian year 2020, those production members will work to create a dark, oppressive, and dirty-looking future.

It's important to quickly communicate any problems or concerns that you anticipate or notice, because problems can cause expensive schedule delays. For example, if a set is being designed that cannot accommodate the lighting equipment that will be needed, the set designer needs to know as soon as possible. Expect that there will be conflicts from time to time and devise an effective way to resolve conflicts constructively. Most conflicts will be resolved with some degree of compromise on all sides; expect compromises to be somewhat more challenging on productions with smaller budgets and teams.

Consider what communication tools will best coordinate a particular team. Some teams may be accustomed to using specific messaging applications; others may use email or text messages. Because teams are often assembled for specific productions, be adaptable if your new team communicates in a different way than you're used to.

Video production teams tend to be temporary and fluid. Even on a television series, the director, screenwriter, and other crew can change from episode to episode. This is a major reason that communication and conflict resolution skills are so important. Video production is a field where teams are hired on word-of-mouth recommendations from the top down. The director will call production supervisors he or she has enjoyed working with in the past, and each supervisor will recommend the crew that they feel they can depend on. Who doesn't get called or recommended? People who had difficulty with team communication or resolving issues, allowing problems to grow and productions to fall behind schedule. You want to act in a way that makes you an attentive, productive, and communicative team member so that those you worked with will recommend you when future projects come up.

★ *ACA Objective 1.5* # Visual Standards and Techniques

Video and film production have evolved a standard vocabulary for visual storytelling. Part of this vocabulary involves how you set up the shots for each scene, which are described in this section. Another part of the vocabulary involves how you edit the shots you captured.

Types of shots

At first glance, a video frame is a simple two-dimensional image. But in video and film production, there's much more to it than that. As a still frame, a video frame is like a photograph in that it's a two-dimensional projection of a three-dimensional space. That means you can use techniques such as perspective to alter how the audience interprets the implied three-dimensional space.

But video isn't just photography because it isn't a still medium. A video frame can contain motion, so you can control how elements of the frame change over time to manipulate the perception of space even more.

USING DISTANCE AND FIELD OF VIEW

These are the types of shot you'll use to tell your story with video,

- **Wide shot, long shot, establishing shot, master shot.** This category of shots typically uses a wide-angle lens to show much more than the subject of the shot, including a lot of the location context around the subject. That's why a wide or long shot is often used as an establishing shot at the beginning of a scene; it tells the audience that the location or time has changed and indicates where and when the new scene is set. If the establishing shot shows an old-looking city with the Eiffel Tower and horse-drawn carriages on the street, you know the action is probably taking place in late 19th-century Paris.

    In project 3, the video clip wideshot.mp4 is an example of a wide establishing shot. While it shows the subject, most of the frame is devoted to the setting so that you understand that the subject is in a school lunchroom (**Figure 6.1**).

- **Tight shot or close-up.** The opposite of the wide shot, a tight shot uses a lens with a narrow field of view to fill the frame with a detail, such as a character's face. A tight shot is typically used to focus the audience's attention on something that would otherwise be difficult to notice in a more normal shot. It's often used to clearly show the facial expressions of an actor or to heighten the effect of a particular line of dialogue. A close-up can be used as a reaction shot, which shows how a character emotionally reacts to action in the scene.

Figure 6.1 Wide shot

Figure 6.2 Tight shot

In project 3, the video clip s4c.mp4 is an example of a tight shot. The frame is dominated by the subject's head (**Figure 6.2**), with so little background visible that it is not clear from this shot where he is (his location is made clear by the previous shots).

- **Mid-shot.** This shot is frequently used because it most closely resembles what is seen with normal vision, between the extremes of the wide shot and tight shot. It doesn't emphasize the subject or the background, so the audience can gain information from both.
- **Two-shot.** When two people are talking, the frame can be composed around them (**Figure 6.3**). One way that it's used is when the director would like the audience to see the expressions of both actors at the same time.

Figure 6.3 Two shot

- **Zoom shot.** Instead of using a single focal length, a zoom shot can use a zoom lens to start at one focal length and end at another. A wide or medium shot can quickly become a close-up without needing an edit, and vice versa.

But because most scenes are built using fixed focal length lenses and edits, the zoom shot is so unexpectedly dynamic that it can be distracting as an effect. Used well, it can convey a sudden change in perspective or urgency.

In project 3, the video clip wideshot.mp4 is a zoom shot. It starts wide but zooms into a medium close-up (**Figure 6.4**).

Figure 6.4 Zoom shot

DIRECTING ATTENTION

Some shots use camera motion or camera adjustments to alter how the audience perceives distance and space in the frame:

- **Deep focus shot.** In a deep focus shot, everything is sharp, no matter how close or far it is (**Figure 6.5**). This is called having a wide **depth of field**. A shot can use deep focus when the director would like the audience to clearly see everything in the frame because important action is happening both near the camera and far from it. Deep focus can be produced by a small lens aperture but is a natural attribute of a wide-angle lens at most apertures.

- **Shallow focus shot.** A shallow focus shot has narrow depth of field, so the frame is in focus only within a narrow range of distance from the camera. Like a tight shot, a shallow focus shot helps focus audience attention on the only content in the frame that is clearly shown (**Figure 6.6**).

Figure 6.5 Deep focus shot   **Figure 6.6** Shallow focus shot

- **Rack focus shot.** A rack focus shot starts focused at one distance and ends focused on a different distance. Rack focusing can be effective for directing audience attention to different parts of the frame. When it's also a shallow focus shot, where everything except the focus distance is completely blurry, a rack focus shot can also be a way of concealing a second subject in the shot and revealing it only by changing the focus distance. For example, a shot can start with a foreground character talking on the phone with a blurred background, and when the focus is changed to the background, the foreground character is blurred but the background now sharply reveals a second character who was overhearing the phone conversation (**Figure 6.7**).

Figure 6.7 A camera set to shallow focus is first focused on the woman in the foreground and then on the man in the background.

- **Dolly shot.** A dolly is a cart used to move a camera during a shot, typically along tracks. A dolly shot is one where the camera moves toward or away from the subject. In a tracking shot, which can also use a dolly, the camera moves along with the character—for example, as the character walks down a street. Both types of shots are used so that the camera does not have to be locked down in one place (**Figure 6.8**).

Figure 6.8 The camera is pushed forward past the woman toward the man.

- **Pan shot.** In a pan shot, the camera is locked down in one place but rotated so that it can capture a panoramic view of a location or set. Though most are horizontal, a vertical pan shot is also possible when, for example, the camera pans up a tall building from the ground. While a pan shot is often used to show a view that is much larger than the lens can take in during a single shot, it can also be used to convey large size or to reveal something that is originally outside the frame.

  In Project 1, the video clip tiltUp.mp4 is a vertical pan shot. It starts with graded dirt and pans up to reveal construction equipment.

TAKING A POINT OF VIEW

The camera is often positioned as an observer at a certain eye-level distance from the subject. But it's also possible to tell the story using the camera to take different points of view, including the point of view of another character:

- **High angle/bird's-eye shot.** A high-angle shot is sometimes called a bird's-eye shot because the camera is looking down on the subject. It is naturally used when one character is talking to another from a higher level, such as from a balcony to a street. A high-angle shot can also be used to make the subject seem small or vulnerable. Some establishing shots can use a high angle combined with wide angle.

  In Project 1, many of the clips are high-angle shots because they were captured using a camera on a drone (**Figure 6.9**). In the past, before drones were practical and affordable, high-angle shots had to be done with cranes, so they are also called *crane shots*.

Figure 6.9
High-angle shot

- **Low-angle/worm's-eye shot.** A low-angle shot is sometimes called a worm's-eye shot because the camera looks up at the subject from below, typically at ground level, but sometimes even below ground depending on the script. In terms of storytelling, a low-angle shot can imply that a subject is massive, powerful, or intimidating, or simply emphasize bulk and volume. In Chapter 3, one of the shot choices is a low-angle shot to emphasize the jump over the table.

   In Project 3, the shot of the jumping student is a low-angle shot (**Figure 6.10**).

Figure 6.10
Low-angle shot

- **Reverse-angle shot.** In a reverse shot, the camera is placed 180 degrees from where it was in the previous shot (**Figure 6.11**). A common example is two characters conversing while facing each other, where the shots alternate between the points of view of each character. Another example would be a close-up shot straight on to the face of a character vividly reacting to what he is seeing, and the next shot from the point of view of the character showing what he's reacting to.

Figure 6.11 Reverse-angle over-the-shoulder shots

- **Over-the-shoulder shot.** Commonly used for conversations, and often used for reaction shots and reverse-angle shots, the over-the-shoulder shot uses the head and shoulders of one character to frame the view of a second character. Having part of the first character in the frame helps orient the audience spatially.

Composing the video frame

If you are already a still photographer or visual artist, you have experience in composing a two-dimensional frame. You can apply the same principles to video, but keep an eye on shots where composition is affected by action during a shot. The difference is that still frame composition is static, whereas video frame composition can be dynamic—the frame content can change over time. This is not a bad thing; it's an additional opportunity for creative expression and storytelling if you can master it. To the audience, a shot can be immensely satisfying as the frame composition continues to work at any time during a shot, even as the camera or the subject moves. The appeal of great dynamic composition holds true for both large battle scenes and more intimate scenes where characters talk while walking.

The principles of composition include the following:

- **Framing.** You can use props or other elements of a set or location to occupy parts of the frame, so as to draw attention to a subject in the remaining part of the frame. In other words, you create a sort of secondary frame (**Figure 6.12**).

Figure 6.12 The subject is framed in the space between the woman's arm and body.

- **Symmetry.** In a symmetrical composition, the composition appears to be reflected across the frame's horizontal or vertical axis (**Figure 6.13**). Symmetry can convey a sense of a static situation, stability, or rigidity. But symmetry can also be used to imply opposition.

Figure 6.13 A symmetrical composition

- **Balance.** In a balanced composition, the visual weight of various elements conveys an equilibrium, while the composition doesn't have to be precisely symmetrical (**Figure 6.14**). Sometimes, how areas of light, shadow, and color are positioned in the frame creates balance regardless of the positions of actual objects.

Figure 6.14 A balanced asymmetrical composition

- **Lead room.** When a subject faces a direction, it's often good to leave more space, or lead room, on the side that they face. Leaving lead room actually achieves a form of dynamic balance that's more effective than centering the subject in the frame (**Figure 6.15**). This is especially true if the subject is moving, because lead room mentally provides the moving subject room to move, even if the subject will never actually reach the edge of the frame during the shot.

- **Leading lines.** You can use converging or angled lines in a frame to help direct audience attention to the subject (**Figure 6.16**). Natural lines of linear perspective are often used for this, such as the lines created by buildings.

Figure 6.15 Lead room is intentionally included on the side of the frame that the subject is facing.

Figure 6.16 The leading lines of the door handles and ceiling tiles point to where the subject appears after he runs around the corner.

- **Rule of thirds.** A classic rule of composition is to divide a frame into horizontal and vertical thirds and place subjects at the intersections of the divisions. This is a quick way to achieve simple nonsymmetrical balance. But keep an eye out for times when the rule doesn't produce an effective result; some shots may work better with another compositional technique.

    One of the reasons Figure 6.14 appears balanced is that the frame was composed so that the bulldozer would pass through the intersection of the lines dividing the top and left thirds of the frame. Figure 6.15 places the head of its subject over the intersection of the lines dividing the top and right thirds of the frame.

- **Watch out for distractions.** Especially on location, objects not specified in the script may distract from the action. For example, if you are shooting subjects standing on a beach in front of the sea, the frame is relatively uncomplicated, using just a few colors. But if someone wearing a bright red swimsuit walks to the water far in the background, that small spot of bright color may distract from the action and ruin the shot. Always remember to look over the entire frame.

- **180-degree rule.** An audience orients itself in a scene by noticing the relative placement of characters and objects in the frame. For example, in a conversation, one character is typically on the left side of the frame and a second character is on the right. If the camera perspective is changed during the scene, it's best for the scene to include only shots that maintain the general left/right positions of the two characters. If a shot is used where the first character is now on the right and the second is on the left, the audience may become disoriented and be distracted from the story.

This concept is expressed as the 180-degree rule. If you think of a circle around the two characters, representing all of the possible camera angles, the spatial relationship of the subjects in the frame is consistent only if the camera stays on one side of that circle, or 180 degrees of that 360-degree circle. The circle is cut in half using an imaginary line drawn through the two characters, so in cinematography you don't want to "cross the line" or "cross the 180" (**Figure 6.17**).

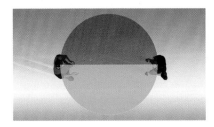

Figure 6.17 As long as the camera stays within the 180-degree half circle on one side of the actors (marked here in green), actors will be oriented consistently from the audience point of view.

If you feel that you need additional training in frame composition, the time-tested advice is to study art history. The masters of painting, drawing, and photography understood the power of composition and spent much of their lives studying and perfecting compositional techniques. Artists of all disciplines draw upon this work.

★ ACA Objective 1.5

Editing techniques

As there are standard techniques for communicating a story through shot selection, there is also a standard vocabulary used when editing the shots in postproduction.

CONTINUITY EDITING

A high priority in video editing is maintaining continuity—maintaining consistency in content and composition so that the audience can easily believe that action after a cut follows logically from the action before a cut. Good continuity makes it easier for an audience to follow a story. Here are some continuity techniques:

- **Match on action.** The action on both sides of a cut matches up. In other words, action in progress at a cut is continued in the next clip.
- **Match cut.** A match cut is less about action and more about content composition. In a match cut, the frames before and after a cut have similar compositions. For example, the last frames of a flashback scene may show a steering wheel being driven aggressively, which cuts to an identically composed shot of the spinning wheel of a car that has crashed and overturned, which turns out to be the same car. The similar composition at the cut can help tie together scenes that are otherwise separated by location or time.
- **Eyeline match.** A shot shows a character looking at something not in the frame, and after the cut is a shot showing what they were looking at but without the character (**Figure 6.18**). The gaze ties the two shots together for the audience, even though the two shots may have no content in common.

Another version of the eyeline match is when one shot shows a character looking at a subject off screen, and another shot shows another character looking at the same off-screen subject. If the two shots are consistent, the audience knows both characters are looking at the same subject.

Figure 6.18 The student looks down at an off-screen object, which turns out to be his watch.

- **Cutaway shots and B-roll.** Although covering the main action is the highest priority, other types of shots can help maintain continuity and provide context. *Cutaway* shots can help establish what is going on around the subject; if the subject is in a war zone, the script may specify cutaway shots of soldiers aiming weapons, medics carrying wounded, and fighter planes flying overhead. In Chapter 2, you used *B-roll* footage of snowboarding to supplement the interview.

  These types of secondary shots can function as buffer shots when they need to be used to help patch over minor discontinuities in the main footage. For example, if it is decided in postproduction that it's confusing that a character is shown in one city and then in a visually different city in the next scene, they may decide to insert a clip of a jetliner landing to imply that a journey has taken place, creating smoother continuity. When planned from the beginning, this type of shot is called a bridging shot.

- **Persistent audio across cuts.** Sound designers often record ambient sound of a location that will play in the background through all of the shots of a scene, to help the audience perceive that sequence as a single scene no matter how many visual cuts there are. The L-cut and J-cut that you learned are similar ways of using one clip's audio to provide better continuity across multiple shots.

MANIPULATING TIME

While continuity editing is a requirement for straight storytelling and nonfiction programs, some editing techniques that treat time in a less linear fashion can enhance more creative projects.

- **Cross-cut.** Cross-cutting is when the editor alternates shots from different events happening at about the same time, instead of a sequence of shots that are all at the same place and time. This is often used to contrast the actions of two subjects, such as two boxers training to fight each other, or a romantic couple preparing for a wedding with each of their families. The comparison can emphasize differences or imply similarities and relationships.

- **Flashback.** Instead of scenes progressing forward in time, one or more scenes depict a past event because it informs the storyline, often shedding light on a character's nature or motivation.

- **Flash forward.** In a flash forward, a scene is inserted that depicts an event in the future, to provide insight into what is happening in the storyline. A flash forward sometimes depicts a future in the storyline that is only possible or imagined, depending on what the characters do next. But if it's made clear that the flash forward is the actual future in the storyline, it then creates tension through the mystery of how the current characters end up with that destiny.

- **Jump cut.** A jump cut creates an intentional discontinuity that's noticed because of an obvious and sharp break in the positions of the camera or the subjects. It can communicate time passing by a short amount and can also convey impatience.
- **Bridging.** The bridging shot discussed earlier is a way of communicating passages of time that are skipped over in the script, such as days or years going by.

MANIPULATING COLOR

Color grading is a valuable tool for helping to differentiate scenes. Scenes in the past may be graded to be faded or sepia-toned; gritty scenes may be graded with more contrast and less saturation. Night scenes may have a blue color cast and deep shadows.

The color controls and adjustment layers you tried in Chapter 3 make it easy to assign a specific look to each location and time period in a production. When you use these consistently, they reinforce place and time for the audience. This can be especially helpful when cross-cutting.

Breaking the rules

The techniques and rules of video production are intended to give the audience consistent and meaningful visual and aural cues to orient them in the space and time of the motion picture frame. Because audiences are accustomed to these rules, it is also useful to break the rules or use them in unconventional ways when a script calls for something other than conventional storytelling. For example:

- **Intentional discontinuity.** Instead of providing the audience with a step-by-step progression through a timeline of recognizable locations, a script may omit establishing shots and bridging shots, use flashbacks and flash forwards, and providing incomplete context for scenes and characters so that the audience becomes curious about the missing pieces. Naturally, breaking the conventions of normal storytelling is a necessary tool in some genres such as mystery stories, where the entire point is to keep the audience guessing at the solution that makes sense of the confusion and ties together all the clues.

One example of this would be an opening shot of a shadowy figure turning a valve in a dark, apparently underground room. What is he doing, where is he, and what will be the consequence of his apparently clandestine activity? The mystery created by the missing context hopefully creates enough curiosity in an audience to motivate them to watch the rest of the program.

- **Extreme focal lengths or perspectives.** Fisheye lenses, extreme close-ups, and odd angles can increase intensity, heighten suspense, and create intentional disorientation.
- **Unbalanced compositions.** Because balanced compositions feel grounded and steady, intentionally unbalanced compositions can create a sense of unease and tension.
- **Dutch angle.** Intentionally tilting the camera for an entire shot can convey unease and suspense, because the conventional practice is to level the camera for every shot. The deviation from the standard creates a sense of discomfort and anxiety.

The techniques in this section are only a sampling of the full range of cinematic techniques that are available. And techniques evolve over time as visual storytelling evolves and the capabilities of our tools change. If you pursue further study in video production, you will likely learn many more ways to tell your story through video.

Licensing, Rights, and Releases

 ACA Objective 1.3

The easy availability of images, video, and music on the Internet has led to an extremely creative remix culture, where people can download almost any type of content they can find and incorporate that content into their own work. However, the media industry has a structure of intellectual property rights, licenses, and permissions that has a history that not only goes back decades but is also backed up by laws and law enforcement. Although it's fun and easy to play with any media you can find, once you step into the video industry you must respect the intellectual property rights connected to any media you use. If you don't, you may expose you and your clients to legal action that can result in large fines and other potential penalties. You can safeguard your career in the video industry by clearly understanding intellectual property rights and abiding by them.

NOTE

The information in this section should not be construed as legal advice. Consult an attorney to better understand the laws specific to the jurisdiction where you live and do business.

Types of licenses

It's very important to understand the kind of license that comes with a media item you download and then use in a project. If you misunderstand the terms of use for an item, legal issues may result. If you create a video for a company or organization and you use a downloaded media item improperly, that company or organization may also be exposed to legal trouble, and that won't be good for your reputation. So be diligent about understanding and abiding by media licenses. You may think that because a project is small and noncommercial no one will notice or care, but you never know when a video will go viral. And keep in mind that the discovery of unauthorized media use is increasingly automated; media companies use software to routinely search uploaded videos for content violations. It's best to keep your legal bases covered at all times.

Copyrighted material requires permission for use. Attribution is not sufficient. In the United States, copyright is automatically assigned to creators at the time they create a work, and it stays with them unless it's signed away in writing. If a work is also registered with the US Copyright Office, the creator may be able to win a higher financial judgment against an infringer (someone who uses the work without permission).

You might be able to use a small portion of a copyrighted work without permission if it falls within the definition of **fair use**. Fair use is a legal term with a specific definition, so research and understand the definition of fair use in your country before assuming you can claim fair use for the way you plan to use a copyrighted work in your project.

Some media comes with no restrictions because it is considered in the **public domain**. There is a common misconception that if the public can easily download a certain media file like a picture or a song, it must be public domain. Legally, that is incorrect. Public domain doesn't depend on how available something is; it's a specific legal term with specific requirements that can differ in various countries. If you see something on the Internet that you'd like to use, you can't assume it's public domain unless it's marked as such, so always check the license or contact the creator for permission.

You might hear people refer to a **Creative Commons** license. Creative Commons is a set of multiple license types that offer variations between copyright and public domain. It's meant to give you more ways to strike a better balance between copyright (which requires permission) and public domain (where the creator has no ownership or control). When you choose a Creative Commons license, be aware of which one you are choosing and why, and be sure to abide by its terms. For example, if you choose a Creative Commons Attribution license, you may, among other things, "copy and redistribute the material in any medium or format" but you "must give appropriate credit, provide a link to the license, and indicate if changes were made."

To read more about the different types of Creative Commons licenses, visit:

https://creativecommons.org/share-your-work/licensing-types-examples/

Free and inexpensive licenses are typically **nonexclusive**, meaning the media is open for anyone to use. A high-profile client such as a major brand may be willing to pay more for an **exclusive** license, which can restrict others from using the same media. For example, a luxury car brand wanting to use a specific song in a commercial may get a license granting exclusivity for five years so that during that time the same song won't also be used in commercials promoting a used car dealership or cat litter.

Obtaining model and property releases

The purpose of a **model release** is to legally affirm that you have permission to depict the people in your video. It's called a release because it releases you from legal liability. The example in this chapter is a school production that involves people under 18 years of age. Because they are minors, it is necessary to obtain signed model releases from their parents.

You can download numerous model release templates from the Internet, and mobile apps are available that claim to be able to generate a legal model release.

If you record video that includes private property, it's safest to obtain a **property release**. As with a model release, a property release legally affirms that you have permission to record imagery of the exterior or interior of a building. It might not be necessary in some cases where the property is not the focus of the shot, such as a small building included as part of a skyline shot including hundreds of buildings, but it's best to consult an attorney familiar with the laws that apply in your area.

Consider legal help

If you build a growing video production business, you can help protect your company by consulting an attorney who is familiar with media-related legal issues. A media attorney can help you understand licenses for media items you want to use, as well as licenses for your own media that you want to distribute or sell. He or she can also review your model release to make sure it's legally valid according to the laws of your country and state.

Moving Into the Industry

You can work in the video industry as an employee or by starting your own company. There are advantages and disadvantages to both ways:

- **Working for a company.** You can get a job working for a video production company, for a television station, or in a video production department of a business, school, or government agency. This type of work has the advantage of being relatively stable, with a regular salary and probably benefits as well. However, depending on where you work, the range of projects may be more limited than you might like, or they may not be the best match to your interests.

- **Working for yourself.** You can start your own video production business. You'll be free to pursue any niche that you are passionate about, to create any kind of project you like, and to accept or reject any project that comes your way. But you will have sole responsibility for finding jobs and keeping your business healthy enough to support you and your family. Running your own video business works best if you possess or are willing to develop strong business and marketing skills and have the ability to outsource jobs such as accounting so that you have more time to be creative.

You can try it both ways to see which path is a better fit for your personality and interests. Whichever way you decide to go, your next step is the same: develop a reel of your work, and post it online so that you can share it with those you would like to work with. Show only your best work, and show only work that represents the kind of jobs you want to get. And keep in mind that all video professionals are constantly updating their reels, so it's the same for you; every time you complete more work that you're proud of, you can improve your reel by adding that work to it.

CHAPTER OBJECTIVES

Chapter Learning Objectives

- Extend Premiere Pro with Adobe Creative Cloud desktop apps.
- Extend Premiere Pro with Adobe Creative Cloud mobile apps.
- Identify useful web resources for Premiere Pro.

CHAPTER 7

Wrapping It Up!

You've almost made it to the end of this book—and by now you should be pretty good at using Adobe Premiere Pro CC to create video sequences targeting a variety of specifications and formats. As you've seen through the projects we've worked on, you can use Premiere Pro to create a diverse array of project types.

Premiere Pro is part of a larger family of applications and services that support it, so before we close, let's tie things up with a look at that big picture, including some resources that you can use to continue exploring Premiere Pro on your own.

Extending Premiere Pro CC with Adobe Creative Cloud

One of the first things you learned in this book is that video production is often a team effort. Although some small productions can be achieved by a single person, when quality is a priority, it quickly becomes necessary to work with specialists who have a deeper knowledge of specific disciplines such as audio recording and editing, still graphics, titles, and color grading.

That kind of deep specialization also extends to the software you use. Even if you can work on a project by yourself, you may find that some phases of production can benefit from more specialized software. Adobe has a strong lineup of desktop applications that can handle many aspects of video production, and Adobe mobile applications are becoming increasingly useful as well. If you have a full subscription to Adobe Creative Cloud (not a Premiere Pro–only subscription), it's a good idea to be aware of these resources because all of them are already included with your

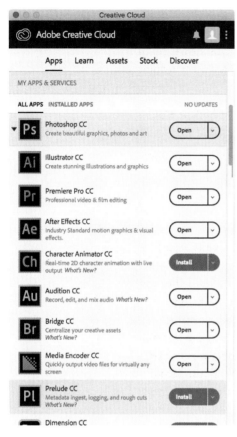

Figure 7.1 You can install more applications using the Adobe Creative Cloud desktop application.

subscription. You can install many of the desktop applications in this chapter using the same Adobe Creative Cloud desktop application (**Figure 7.1**) that you used to install Premiere Pro, and you can find out more about them at the Adobe Creative Cloud applications page (www.adobe.com/creativecloud/catalog/desktop.html).

Video production applications

Many Adobe Creative Cloud desktop applications support various aspects of video production. You may not need them all the time, but when you do, it's good to know that they're immediately accessible.

STREAMLINING CLIP LOGGING AND ROUGH CUTS WITH ADOBE PRELUDE CC

Adobe Prelude can simplify metadata entry, logging, and rough-cut creation. It can be a useful tool for producers or directors who want to perform preproduction tasks before handing off clips for editing in Premiere Pro.

Do you remember how the Export Media dialog box lets you enter metadata such as keywords and rights information for the video you're exporting? Metadata can be even more useful when you enter it as you log video clips in preproduction. Many productions like to enter clip metadata such as scene numbers, actors, and notes, and Prelude is a simple way to do this.

You can even do your rough cut in Prelude. When you're ready to refine your rough cut, you can send it directly to Premiere Pro, where you can continue editing it using the full range of editing tools in Premiere Pro.

To find out more, go to the Adobe Prelude product page (www.adobe.com/products/prelude.html).

EDITING AND MIXING AUDIO WITH ADOBE AUDITION CC

Adobe Audition is a professional digital audio workstation application; you can think of it as the audio equivalent of Premiere Pro. You can turn to Audition to fix audio problems that are too challenging for Premiere Pro. For example, you can use

Audition to clean up audio by removing noise and unwanted sounds such as pops. Doing so is easy because of the integration between Audition and Premiere Pro. You can jump to Audition from an audio clip selected in a Premiere Pro sequence, edit it, and return to Premiere Pro with the edited clip in place—without having to manually export or import files.

You can also use Audition for mixing video soundtracks using specialized audio tools that are more powerful and versatile than those available in Premiere Pro.

To learn more about Audition, go to the Audition product page (www.adobe.com/products/audition.html).

MAKING EXPRESSIVE ANIMATIONS WITH ADOBE CHARACTER ANIMATOR CC

If you're creating a production using animated characters, such as a cartoon, a children's program, or an educational video, you might be wondering how much work it will take to animate your characters. Adobe Character Animator makes it fast and easy; it uses your computer's camera and microphone to track your head movements and facial expressions and applies those to the animated character in real time. You can also use Character Animator to move a character's arms and legs. You can use one of the animated puppets that's included with Character Animator, or you can build your own characters using Photoshop or Illustrator. You can record your animations to video files and then add them to a Premiere Pro project and edit them into a complete program.

To learn more, go to the Character Animator product page (www.adobe.com/products/character-animator.html).

CREATING HOLLYWOOD-STYLE SPECIAL EFFECTS WITH ADOBE AFTER EFFECTS CC

You learned how to make clips move, rotate, and scale by using keyframes to animate settings such as Position, Rotation, and Scale in the Effect Controls panel in Premiere Pro. And you learned how to composite clips on different tracks using masks. But some productions require special effects at a level of detail and quality that's beyond what Premiere Pro can achieve, and when you need that, you can call on Adobe After Effects.

With After Effects you can produce cinematic visual effects. You have more precise, powerful, and flexible control over features such as masking and keyframes than in Premiere Pro, and you can also create animations and special effects in 2D and 3D.

After Effects has features such as motion tracking (tying the animation of one layer to the motion of another) and rotoscoping (automatically tracking changes in a mask from frame to frame). These features make it easy to place actors in virtual sets or create animated fly-through logos. If you want to create the next great science fiction epic, After Effects will probably be a big part of your workflow.

Of course, the sophisticated effects shots you put together in After Effects can be imported into Premiere Pro and edited with other clips. As with Audition, After Effects and Premiere Pro can exchange data directly so you don't have to export and import files.

To learn more, go to the After Effects product page (www.adobe.com/products/aftereffects.html).

CREATING OUTPUT EFFICIENTLY WITH ADOBE MEDIA ENCODER CC

In this book you've seen how Adobe Media Encoder can increase your productivity. By sending final video from Premiere Pro to Media Encoder for rendering in the background, you free up Premiere Pro so you can continue working on another sequence. You don't have to stand by Premiere Pro watching for a job to finish just so you can manually start another one, because you can simply add multiple items to the Media Encoder queue and have it render each job in turn, unattended, until all of them are done. And if you need to create versions of the same sequence for different output formats, just duplicate it in Media Encoder and apply different export presets. After you start the queue, you can get some sleep or go out while it renders all the export jobs for you.

Adobe Media Encoder is useful for more than just rendering edited sequences in the background. You can also use Media Encoder to transcode source clips (re-encoding them into a different format), create proxies of source clips (as you learned in Chapter 3), and create video-only or (as you learned in Chapter 4) audio-only versions of a sequence. To learn more, go to the Adobe Media Encoder product page: https://www.adobe.com/products/media-encoder.html.

FINDING, ORGANIZING, AND INSPECTING MEDIA WITH ADOBE BRIDGE

While you can browse and inspect information about media using the Project panel and Media Browser panel in Premiere Pro, Adobe Bridge is also a great standalone tool for media browsing and organization, especially for files not yet imported into Premiere Pro and in file formats used by Adobe Creative Cloud applications. You can use Bridge to preview video files and view their specifications and metadata without having to open them.

In Bridge you can use powerful searching and filtering to find files quickly. For example, you can have Bridge show you all the files within a folder and its sub-folders, and then you can use a filter to show you only the video files. Then you can drag those video files directly to the Project panel in Premiere Pro, because drag-and-drop importing works just as well from Bridge as from the folders on your desktop.

Bridge is also a convenient way to view and edit metadata for many file types. For example, you can select hundreds of still images and apply ownership information and keywords to all of the selected images at once. In short, you can think of Bridge as a more powerful version of your operating system desktop, optimized for Creative Cloud applications.

To learn more, go to the Bridge product page (www.adobe.com/products/bridge.html).

Graphics applications

In this book you've imported photographic still images and other graphics into video sequences. Those types of media are typically prepared in two other important Adobe Creative Cloud applications.

EDITING PHOTOGRAPHS WITH ADOBE PHOTOSHOP CC

If there's any application in this chapter that you might have heard of, it's probably Adobe Photoshop, the standard for digital photographic editing and image processing for many years. When you have photographs that need to be adjusted, cleaned up, or composited before being imported into Premiere Pro, Photoshop is the tool to use. Premiere Pro understands the layered Photoshop file format directly, as well as other image file formats such as JPG that Photoshop can also save and export.

Although many photographers shoot in a camera's raw file format, Premiere Pro doesn't read camera raw files directly. Photoshop includes the Adobe Camera Raw plug-in, which you can use along with Photoshop to adjust camera raw format files and save them in an image format that Premiere Pro can import.

To learn more, go to the Photoshop product page (www.adobe.com/products/photoshop.html).

CREATING SMOOTH LINE ART WITH ADOBE ILLUSTRATOR CC

A video production may use nonphotographic art and graphics, such as logos, charts, diagrams, and maps. These types of graphics are often created in a drawing program such as Adobe Illustrator. Whereas Photoshop creates *bitmap graphics*, or pictures made of a grid of pixels, with Illustrator you can create *vector graphics,* or pictures defined by points and paths. Remember when you drew a mask in Chapter 4? When you laid down points to define path segments, you were creating vector paths as you would in Illustrator.

For Premiere Pro users, the big advantage of Illustrator vector graphics is that they scale smoothly to any size. When you enlarge a bitmap graphic such as a photograph, scaling above a certain percentage results in blocky images and jagged-edged shapes because the pixels are too large. Vector graphics enlarge smoothly because the points and paths that define the shapes are recalculated at the current resolution. Because of that, the shapes always use all of the available frame pixels, so they always have a smooth edge.

To learn more, go to the Illustrator product page (www.adobe.com/products/illustrator.html).

Mobile apps

Smartphones and tablets continue to become more powerful as their cameras continue to get better, with some capable of recording and editing 4K video. That combination makes video editing easier to achieve on a mobile device. Adobe has mobile apps that help you do that, and they can link up to Premiere Pro. Even if you're not interested in actually editing video on a mobile device, Adobe also has mobile apps that can support a production in other ways.

- **Adobe Premiere Clip** lets you edit video on a mobile device. You can create a video from media on your device, sync it to music, add titles, and apply slow motion and other effects. You can send a Clip project directly to Premiere Pro for further editing.
- **Adobe Capture** can sample multiple types of media and send them directly to Adobe applications. You can use Capture to convert a photo into a vector graphic that you can use in Premiere Pro.
- **Prelude Live Logger** is an app that lets you use an Apple iPad to log clips as they're being recorded with a tapeless camera. The metadata you log can be transferred to the Adobe Prelude desktop application, which can combine the metadata with their clips and then send them on to Premiere Pro for editing.

- **Adobe Spark Video** can combine video clips, photos, and icons with a text overlay, creating an engaging video. Spark Video is not a general-purpose video editor like Premiere Pro or Premiere Clip; it's a template-based app designed specifically for telling video stories that are optimized for social media.

You can find out more about these apps at the Adobe Creative Cloud mobile apps page (www.adobe.com/creativecloud/catalog/mobile.html).

Most Adobe mobile apps are currently available on iOS, and some are available on Android. Check the system requirements for each app.

Cloud-based services

Adobe Creative Cloud is named that way because it isn't just a collection of desktop and mobile applications. What makes it all work are cloud-based services that tie together the applications, making them much more useful together than they would be individually.

USING ASSETS ACROSS DEVICES AND PROJECTS WITH CREATIVE CLOUD LIBRARIES

Creative Cloud Libraries are cloud-based storage of graphics and other assets that automatically sync among Creative Cloud desktop and mobile applications. For example, if you capture a color Look with the Adobe Capture mobile application, it will sync with Creative Cloud Libraries so that the next time you open Premiere Pro, the Look will appear in the Libraries panel, ready for use in any project.

Because assets in your Libraries panel are cloud-based, they're available to any computer that's signed into your Creative Cloud account. For example, if you added assets to your Libraries panel while you were working in the field on your laptop, when you return to your desktop and start Premiere Pro, those assets will appear in the Libraries panel on that computer too.

EASILY ACQUIRING AND LICENSING ASSETS WITH ADOBE STOCK

There are occasions when you don't have the time, budget, or resources to acquire every asset you need for a production. For example, you may need video of lightning in the sky but there isn't going to be a storm in your area before your project's deadline. That's when stock assets are useful. Adobe Stock offers media such as video, images, 3D objects, titles, and motion graphics templates.

While there are many websites offering stock assets, Adobe Stock is well integrated with Adobe applications. For example, the Libraries panel in Premiere Pro has a Search Adobe Stock button so that you can look for stock media without leaving Premiere Pro. If you find an asset you like, you can save a preview, which appears in the Libraries panel so you can try it before you pay for it. If you decide to keep it, you can license it directly from the Libraries panel. If you used another stock website, you'd have to manually download and import the stock asset; with Adobe Stock, the integration with the Libraries panel means that as soon as you license the asset, it's already available inside Premiere Pro as soon as it finishes downloading.

A small number of Adobe Stock assets are free, such as some of the title templates available in the Essential Graphics panel in Premiere Pro. But access to the full range of Adobe Stock assets requires a plan that in most cases is a separate cost that you add on to your Creative Cloud plan. To see which Adobe Stock plans are available to you, go to the Adobe Stock website (stock.adobe.com).

SHOWING OFF YOUR WORK WITH ADOBE PORTFOLIO

If you want to create a web page containing your best work, you can use the Adobe Portfolio website building service that's included with a paid Creative Cloud membership and includes hosting. Creating your portfolio website can be as easy as picking a template and filling it in with your own text and content, which can include images and embedded video. For example, if you have videos spread across multiple online websites such as YouTube and Vimeo, you can create a web page or gallery that has all of your videos in one place. If you want to customize your website, Portfolio provides many settings for changing fonts, colors, margins, backgrounds, and other design options. If you have a domain name, you can set it as the web address of your Adobe Portfolio site.

To learn more, go to the Adobe Portfolio website (myportfolio.com).

OPTIMIZING STUDIO WORKFLOWS WITH ADOBE TEAM PROJECTS

If you work with a group of people at a busy video production studio, whether independently or as part of a larger organization such as a corporation or a television station, you might face challenges coordinating the flow of video, video metadata, and other files among members of your team in the office and across multiple locations.

Adobe Team Projects is made for those kinds of organizations; for one thing, it's available only to the Creative Cloud for Enterprise and Creative Cloud for Teams plans. Team Projects works with Adobe Premiere Pro CC, After Effects CC, and Prelude CC. It's cloud-hosted, so members of a workgroup can edit video from their desks over the network instead of having to carry storage drives from person to person. It also has features such as version control that help the team know where specific files are in the production workflow and who is currently using them. In this way, Adobe Team Projects helps team members work more efficiently by reducing the amount of file transfer and file management they have to do.

If you're part of an organization that might benefit from Adobe Team Projects, read more about it at the Team Projects product page (helpx.adobe.com/premiere-pro/how-to/team-project.html).

Where to Go Next

Adobe has several web-based resources that are designed to serve the Premiere Pro community, including the following:

- **Premiere Pro CC product page** (www.adobe.com/products/premiere.html): This is the main page for Premiere Pro, with links to product information, tutorials, and help resources.

- **Learn Premiere Pro CC** (helpx.adobe.com/premiere-pro/tutorials.html): Learn more about Premiere Pro by watching video demonstrations of features and techniques. This is also a great way to catch up when new features are added.

- **Main Premiere Pro CC support page** (helpx.adobe.com/support/premiere-pro.html): This page gathers links to a range of support resources on one page, including tutorials, articles, videos, common questions and top issues, community forums, and customer support contact information.

- **Premiere Pro CC help and online user guide** (helpx.adobe.com/premiere-pro/user-guide.html): Browse or search the online user guide for Premiere Pro for descriptions of tools, commands, effects, and other features.

- **Premiere Pro CC community** (forums.adobe.com/community/premiere): Ask questions about Premiere Pro at Adobe Communities, where questions are answered by other community members.

- **Premiere Pro CC blog** (theblog.adobe.com/creative-cloud/premiere-pro/): As a Creative Cloud application, Premiere Pro may be updated with new features at any time. Posts on the Premiere Pro blog can keep you informed about changes and new features when you see a Premiere Pro update in the Creative Cloud desktop application. You may also see posts about previews of upcoming technologies and other developments you may want to be aware of.

Good Luck, and Have Fun!

Congratulations! You've completed all the chapters in this book, and in this chapter you have explored some of the other creative capabilities and resources available to you in Adobe Creative Cloud. Keep practicing with your own projects so that the basics of editing become second nature. Use the resources in this chapter to continue exploring Premiere Pro CC, as you discover your working style and creative voice in video.

ACA Objectives Covered

DOMAIN OBJECTIVES	CHAPTER	VIDEO
DOMAIN 1.0 Working In the Video Industry		
1.1 Identify the purpose, audience, and audience needs for editing video.	**Ch 1** Identifying Job Requirements, 13 **Ch 2** Preproduction, 89 **Ch 3** Getting Ready in Preproduction, 149 **Ch 4** Preproduction, 203 **Ch 5** Preproduction, 223	**2.1** Snowboarding Highlight Video **3.1** Job Requirements **4.1** Introducing the Weather Report Project **5.1** Introducing the Memorial Slide Show Project
1.2 Communicate with colleagues and clients about project plans.	**Ch 1** Identifying Job Requirements, 13 **Ch 2** Preproduction, 89 **Ch 3** Getting Ready in Preproduction, 149 **Ch 4** Preproduction, 203 **Ch 6** Phases of Production, 239 **Ch 6** Roles of a Video Production Team, 246 **Ch 6** Communicating Effectively, 248	**2.1** Snowboarding Highlight Video **3.1** Job Requirements **4.1** Introducing the Weather Report Project
1.3 Determine the type of copyright, permissions, and licensing required to use specific content.	**Ch 1** Identifying Job Requirements, 13 **Ch 3** Acquiring and Creating Media, 151 **Ch 6** Licensing, Rights, and Releases, 263	**3.3** Stock Media
1.4 Demonstrate an understanding of key terminology related to digital video.	**Ch 1** Identifying Job Requirements, 13 **Ch 2** Making Quick Fixes to Color, 110 **Ch 3** Taking Another Look at Preferences, 154 **Ch 3** Inspecting the Properties of a Clip, 163 **Ch 3** Starting a Rough Cut, 164 **Ch 5** Preproduction, 223	**2.9** Basic Color Correction **3.4** Preferences Detailed **3.6** Properties **3.7** Rough Cut **5.1** Introducing the Memorial Slide Show Project
1.5 Demonstrate knowledge of basic design principles and best practices employed in the video industry.	**Ch 2** Applying L and J Cuts, 120 **Ch 6** Visual Standards and Techniques, 250 **Ch 6** Editing Techniques, 260	**2.12** L and J Cuts

continues on next page

continued from previous page

DOMAIN OBJECTIVES	CHAPTER	VIDEO
DOMAIN 2.0 Project Setup and Interface		
2.1 Set appropriate project settings for video.	**Ch 1** Setting Up the New Project Dialog Box, 18 **Ch 2** Setting Up the Interview Project, 90 **Ch 2** Creating the Interview Sequence, 95 **Ch 3** Setting Up the Action Scene Project, 158 **Ch 3** Using Proxies and Removing Unused Clips, 193 **Ch 4** Setting Up a Project, 204 **Ch 5** Setting Up a Slide Show Project, 224 **Ch 5** Creating a Sequence from Multiple Files Quickly, 225	**2.2** Organizing the Media Files **2.3** Setting Up the Project **2.5** Creating a New Sequence **3.17** Proxy and Unused Clips **4.2** Organize Your Project **5.2** Organize Your Project
2.2 Navigate, organize, and customize the application workspace.	**Ch 1** Starting Premiere Pro, 15 **Ch 1** Exploring Panels and Workspaces, 25 **Ch 1** Using Workspaces, 36 **Ch 2** Diving Deeper into the Workspace, 99 **Ch 3** Taking Another Look at Preferences, 154 **Ch 3** Reviewing Timeline Controls, 182	**1.5** Start Premiere Pro **1.6** Exploring Panels and Workspaces **2.6** Diving Deeper into the Workspace **3.4** Preferences Detailed **3.13** Timeline Interface
2.3 Use non-visible design tools in the interface to aid in video workflow.	**Ch 1** Navigating the Timeline, 56 **Ch 1** Navigating the Timeline, 56 **Ch 2** Using Markers, 131 **Ch 3** Reviewing Timeline Controls, 182 **Ch 4** Compositing a Green Screen Clip with a New Background, 207 **Ch 5** Creating a Sequence from Multiple Files Quickly, 225	**1.13** Navigating the Timeline **2.8** Work in the Timeline **2.16** Video Markers **3.13** Timeline Interface **4.4** Key Weatherman over Weathermap
2.4 Import assets into a project.	**Ch 1** Identifying Job Requirements, 14 **Ch 1** Importing Media, 38 **Ch 2** Setting Up the Interview Project, 90 **Ch 3** Setting Up the Action Scene Project, 158 **Ch 3** Importing Files and Maintaining Links, 159 **Ch 3** Inspecting the Properties of a Clip, 163 **Ch 3** Using Proxies and Removing Unused Clips, 193	**1.2** File Management Basics **1.4** Identifying Job Requirements **1.8** Importing Media **2.2** Organizing the Media Files **2.3** Setting Up the Project **3.5** Importing and Linking Media **3.6** Properties **3.17** Proxy and Unused Clips **3.19** Project Manager

DOMAIN OBJECTIVES	CHAPTER	VIDEO
2.4 Import assets into a project. *(continued)*	**Ch 3** Using the Project Manager, 199 **Ch 4** Setting Up a Project, 204 **Ch 4** Adding and Animating More Graphics, 212 **Ch 5** Setting Up a Slide Show Project, 224	**4.2** Organize Your Project **4.5** Add Graphics
DOMAIN 3.0 Organization of Video Projects		
3.1 Use Sequence panel to manage video and audio tracks.	**Ch 1** Editing a Sequence, 44 **Ch 1** Working with Audio, 66 **Ch 2** Playing a Sequence Smoothly, 125 **Ch 3** Nesting Sequences for Different Delivery Requirements, 190 **Ch 3** Using Proxies and Removing Unused Clips, 193 **Ch 5** Creating a Sequence from Multiple Files Quickly, 225	**1.10** Working with Sequences **1.15** Working with Audio **2.14** Render **3.16** Working with Nesting and Special Sequences **3.17** Proxy and Unused Clips
3.2 Modify basic track visibility and audio levels.	**Ch 1** Working with Audio, 66	**1.15** Working with Audio
DOMAIN 4.0 Create and Modify Visual Elements		
4.1 Use core tools and features to edit audio and video.	**Ch 1** Viewing Imported Media, 41 **Ch 1** Understanding a Basic Editing Workflow, 43 **Ch 1** Building a Rough Cut, 49 **Ch 1** Creating a Rough Cut from the Project Panel or Bins, 54 **Ch 1** Exploring the Editing Tools, 57 **Ch 2** Subtracting Unwanted Clip Segments, 114 **Ch 2** Getting Organized in the Timeline Panel, 119 **Ch 3** Starting a Rough Cut, 164 **Ch 3** Editing with Vertical Video, 168 **Ch 3** Editing a Multicam Sequence, 172 **Ch 3** Finishing Sequence Edits, 176	**1.9** Understanding a Basic Editing Workflow **1.11** Create a Rough Cut **1.12** Create a Rough Cut from the Project Panel **1.14** Reviewing the Editing Tools **2.10** Lift and Extract **2.11** Organization in the Timeline **3.7** Rough Cut **3.8** Vertical Video **3.9** Multicam **3.10** Labeling
4.2 Add and manipulate titles using appropriate typographic settings	**Ch 2** Adding Titles, 133 **Ch 3** Adding Credits, 183 **Ch 4** Importing Layered Photoshop Documents, 204	**2.17** Essential Graphics **3.14** Rolling Credits

continues on next page

continued from previous page

DOMAIN OBJECTIVES	CHAPTER	VIDEO
4.3 Trim footage for use in sequences.	**Ch 1** Exploring the Editing Tools, 57 **Ch 1** Adding a Simple Title, 70 **Ch 2** Subtracting Unwanted Clip Segments, 114 **Ch 2** Applying L and J Cuts, 120 **Ch 3** Starting a Rough Cut, 164 **Ch 3** Editing with Vertical Video, 168 **Ch 3** Editing a Multicam Sequence, 172	**1.14** Reviewing the Editing Tools **1.16** Adding a Simple Title **2.10** Lift and Extract **2.12** L and J Cuts **3.7** Rough Cut **3.8** Vertical Video **3.9** Multicam
4.4 Transform digital media within a project.	**Ch 2** Making Quick Fixes to Audio, 105 **Ch 2** Playing a Clip Faster or Slower, 123 **Ch 2** Varying Clip Speed Over Time, 127 **Ch 2** Merging Separate Video and Audio Files, 142	**2.8** Audio Sweetening **2.13** Basic Speed Changes **2.15** Time Remapping **2.19** Merging Clips
4.5 Use basic reconstructing and editing techniques to manipulate digital audio and video.	**Ch 2** Making Quick Fixes to Color, 110 **Ch 2** Stabilizing a Shaky Clip, 139 **Ch 3** Applying an Adjustment Layer, 180 **Ch 4** Setting Up a Project, 204 **Ch 5** Adding a Ken Burns Motion Effect, 230	**2.9** Basic Color Correction **2.18** Stabilize Your video **3.12** Adjustment Layers **4.2** Organize Your Project **5.6** Adding a Ken Burns Motion Effect
4.6 Add and modify effects and transitions.	**Ch 1** Using Video Transitions and Effects, 75 **Ch 2** Stabilizing a Shaky Clip, 139 **Ch 4** Compositing a Green Screen Clip with a New Background, 207 **Ch 4** Adding and Animating More Graphics, 212 **Ch 5** Adding a Ken Burns Motion Effect, 230	**1.17** Using Video Transitions and Effects **2.18** Stabilize Tour Video **4.4** Key Weatherman over Weathermap **4.5** Add Graphics **5.6** Adding a Ken Burns Motion Effect
4.7 Manage audio in a video sequence.	**Ch 1** Working with Audio, 66 **Ch 2** Making Quick Fixes to Audio, 105 **Ch 2** Merging Separate Video and Audio Files, 142 **Ch 3** Starting a Rough Cut, 164 **Ch 3** Editing a Multicam Sequence, 172 **Ch 3** Sweetening Different Audio Types, 177 **Ch 3** Recording a Voiceover, 187	**1.15** Working with Audio **2.8** Audio Sweetening **2.19** Merging Clips **3.7** Rough Cut **3.9** Multicam **3.11** Essential Audio **3.15** Voice Recording

DOMAIN OBJECTIVES	CHAPTER	VIDEO
DOMAIN 5.0 Publishing Digital Media		
5.1 Prepare documents for publishing to web, screen, and other digital devices.	**Ch 1** Exporting a Finished Video File, 80 **Ch 2** Exporting with Adobe Media Encoder CC, 143 **Ch 3** Exporting Multiple Sequences, 197 **Ch 5** Exporting Multiple Versions with Adobe Media Encoder, 232	**1.19** Exporting a Finished Video File **2.15** Clean Up the Timeline and Export Your Project **2.20** Exporting with Adobe Media Encoder **3.18** Exporting Media Encoder Importing Preset **5.76** Export Your Slideshow
5.2 Export a clip, range of frames, or an entire sequence.	**Ch 1** Exporting a Finished Video File, 80 **Ch 2** Exporting with Adobe Media Encoder CC, 143 **Ch 3** Exporting Multiple Sequences, 197 **Ch 5** Exporting Multiple Versions with Adobe Media Encoder, 232	**1.19** Exporting a Finished Video File **2.20** Exporting with Adobe Media Encoder **3.18** Exporting Media Encoder Importing Preset

Glossary

adjustment layer A Premiere Pro media type that carries only effects. You can use an adjustment layer to quickly apply effects to a large number of clips at once simply by adding an adjustment layer to an upper sequence video track; its effects affect all clips on all lower tracks.

alpha channel A channel that carries transparency information for a video clip or still image. An alpha channel is stored with the media in addition to color channels such as RGB.

Apple Metal A computing architecture that allows compatible graphics hardware to accelerate calculations on Apple Mac computers.

aspect ratio The ratio of width to height of a video frame. Not to be confused with pixel aspect ratio.

attribution Crediting the creator of a work that you use in your work. If a license requires attribution, you include the creator in your video credits.

AVCHD (Advanced Video Coding High Definition format) A format for high-definition video recording. An AVCHD clip may exist as a set of files instead of a single file.

B-roll Secondary footage recorded to supplement the primary footage, usually by adding more meaning, context, and depth.

Bezier handles Interactive levers for controlling the direction of curved path segments, named after engineer Pierre Bézier.

bin A container for organizing clips and other media in the Project window. Named after the physical bins that were used historically to organize videotapes.

chroma key compositing Compositing by removing a key color in one clip. The key color is replaced by the contents of another clip on a lower video track, such as an alternate background.

clip A single video or audio source file. In Premiere Pro you import clips and edit them together into a sequence.

compositing Creating a single shot by combining different parts of multiple video clips.

context menu The menu that pops up when you right-click (Windows) or Control-click (macOS). It's called a context menu because it contains commands specific to the object you clicked.

copyright A legal right to reproduce or license a work. If you want to use a work and you do not possess the copyright, you must request permission to use the work.

Creative Commons A set of license types that offer variations between copyright and public domain.

CUDA A computing architecture by NVIDIA that allows compatible NVIDIA graphics hardware to accelerate calculations.

docking (panels) Attaching a panel to another panel, along an edge.

embedding Copying imported content into a document instead of linking to it.

exclusive license A content license that is granted to only one user so that others cannot also license the content. Useful when the user doesn't want the same content showing up in a competitor's work.

exporting Rendering a sequence of video, audio, and other content into a single file.

extract An edit that removes a range of frames from a clip in a sequence and closes the resulting gap with a ripple edit.

fair use A legal principle that allows the use of copyrighted work without requesting permission, in specific cases that vary by region.

floating (panel) A panel or panel group that is not attached to any other panels, existing as its own window.

Forward Delete A key, sometimes labeled Del, that produces a different result than the Delete key in some cases. On extended keyboards it is a discrete key; on smaller keyboards pressing Fn+Delete results in Forward Delete.

frame rate The frequency of video frame playback in frames per second, such as 24 frames per second or 30 frames per second.

garbage matte A rough mask drawn to simplify later compositing steps.

GIF (Graphics Interchange Format) Graphics format for still images and simple animations. Supports a limited 256 color palette and one level of transparency.

green screen A backdrop used for compositing, using a green color that is safe to remove from the entire scene.

grouping (panels) Combining panels so that they share a single space, like a stack of paper folders.

In point For a clip, an In point marks the starting time of the range of frames that will be used in a sequence. For a sequence, an In point marks the starting time of a range of frames for an edit or export.

ingest Transferring captured video clips into a Premiere Pro project, such as from a camera card. Ingest can include steps such as generating proxies and transcoding.

insert An edit that adds a clip to a sequence so that all clips after the insertion time (playhead position) are offset forward in time by the duration of the inserted clip. For example, inserting a two-second clip moves all following clips forward two seconds so that no content is deleted.

J cut An edit where you hear the next video clip's audio before the next video clip appears. The extension of an audio clip under the preceding clip results in a J shape on the timeline.

JPEG (Joint Photographic Experts Group) Graphics format for still images using loss compression. Does not support transparency.

keyframe A timeline marker that allows a setting to vary over time. At least two keyframes are used—one to set the starting value and another to set the ending value.

L cut An edit that extends a video clip's audio into the following video clip, sometimes done to improve continuity.

license Legal permission to use content. A license allows a rights holder to keep their copyright while granting permission for usage.

lift An edit that removes a range of frames from a clip in a sequence, resulting in a gap.

link A filesystem path to a document used in a project. Instead of copying imported content into a Premiere Pro project file, the project file remembers the link. Linking helps keep project file sizes small and makes it easier to manage and update imported content.

lower-third A type of title located in the bottom one-third of a video frame, such as a title displayed to identify a subject or location.

matte Another name for a mask.

Media Browser A panel that is an alternate way of importing media into Premiere Pro. Unlike the Import command, the Media Browser can apply ingest settings and better preserve some multiple-file clip packages such as AVCHD content.

media cache A folder of pre-rendered sequence frames that are retained so that Premiere Pro can quickly play them back, instead of having to recalculate them every time they're played back.

Mercury Playback Engine A set of technologies by Adobe Systems, developed to make video editing faster and more responsive. It takes advantage of multithreaded 64-bit CPU processing, RAM, fast storage, and graphics hardware in whatever combination is available.

metadata Informational data about a source asset. It can be intrinsic properties of a file, such as file size, running time, format, or it can be data added by the user such as keywords or copyright information.

model release A legal release that says the subject of content (such as a person seen in a video clip of photograph) grants permission to the creator to publish the work containing the subject's likeness.

modifier key A key that you hold down to change the function of a feature or keyboard shortcut. For example, pressing the I key sets the In point in the Source or Program panel, but if you press the I key and the Shift key, the Shift key modifies the I key so that it now goes to the In point.

multicam Editing clips from multiple cameras that recorded the same scene at the same time to provide the editor with options such as different angles or focal lengths.

multiplexing How multiple signals are combined into a single signal, such as when audio and video are combined during exporting.

mute Silence an audio track.

nesting Adding a sequence as a clip to another sequence, resulting in one sequence inside another. Nesting can help organize complex programs.

nonexclusive license A content license that allows the creator to license the same work to others.

NTSC National Television System Committee, a technical television standard that was widely used for analog television. It has been largely replaced by the digital ATSC (Advanced Television System Committee) standard.

opacity mask A mask drawn to control the opacity of the region of the frame that it covers.

OpenCL A computing architecture that allows compatible graphics hardware to accelerate calculations.

Out point For a clip, an Out point marks the ending time of the range of frames that will be used in a sequence. For a sequence, an Out point marks the ending time of a range of frames for an edit or export.

overscan Slightly enlarging the video image. For example, overscan can provide a larger image on a small screen. However, overscan pushes the edges of the video image off screen. Overscan is less common today.

overwrite An edit that adds a clip to a sequence so that it replaces any content that already exists within the duration of the clip that's added. For example, if a 3-second clip is added at 10 seconds into a sequence, any clips already on that track between 10 and 13 seconds are replaced by the new clip.

pixel aspect ratio The ratio of width to height of a video pixel. Some video formats use square pixels; some use rectangular pixels. Not to be confused with *aspect ratio*.

playhead Marker that indicates the current time on a time ruler. Also called the current time indicator.

PNG (Portable Network Graphics) Graphics format for still images. Supports many types of compression, along with alpha channels for high-quality transparency.

postproduction Assembling and editing the media assets acquired during production, including the creation of all final versions of the deliverables.

preproduction The planning phase before the production phase of creating a video.

production Recording clips for a video program according to the plan and schedule established in preproduction.

program Another term for a video sequence, which is why the Program panel plays back sequences.

project A Premiere Pro document. A project can contain a list of media used in the project, as well as one or more sequences.

properties Attributes of a source media file, such as its file size, format, and codec.

proxy A low-bitrate placeholder copy of a clip, created to allow smoother playback during editing. When a sequence does not play back smoothly because the bitrate of the original clips is so high that the computer system can't keep up, it can help to generate proxies and use them in place of the original clips while editing. On export, Premiere Pro automatically switches back to using the full-quality original clips.

public domain A legal state where a work is not subject to intellectual property rights and so can be used without requesting permission. The exact legal definition varies by region.

ripple delete Deleting a clip and automatically filling the resulting gap with a ripple edit.

ripple edit An edit that doesn't leave a gap between clips because the time positions of all following clips are adjusted by the amount of the edit.

rolling edit An edit that doesn't leave a gap between clips because the durations of the two clips on both sides of the edit are adjusted by the amount of the edit.

rough cut An early draft of a video program. It validates the approach to editing so that it can be either refined or redone.

royalty-free A type of content license that does not require you to pay a royalty, or fee, for the use of the content.

rubber band A control on a track that lets you adjust a setting by dragging vertically. When keyframes are added to a rubber band, the setting can be varied over time.

safe margins Guides that mark the area of a frame within which text and all of the important action should be composed. Content composed outside the safe margins may be cut off on screens that overscan.

scratch disk A storage location used for various types of audio and video temporary files that are cached or rendered while capturing, importing, or editing video. Each project can have its own scratch disk locations. Scratch disks save the most time when they are both large and very fast.

scrub Adjust a value interactively and visually by dragging the mouse while you watch the results change. You can scrub by dragging the playhead along a time ruler, or you can scrub a setting for an effect such as Opacity by dragging the displayed value.

sequence A series of video clips on one or more tracks. It can include other content such as audio clips and still images. A sequence uses specific video and audio settings, such as 4K video.

slide edit An edit involving a clip that already exists in a sequence, as if it was slid forward or back in time in front of adjacent clips, so that the only changes are to the Out point of the preceding clip and the In point of the following clip.

slip edit An edit involving a clip that already exists in a sequence, where the only changes are to the In point and Out point of the clip, as if it has been slipped behind adjacent clips.

solo Play an audio track alone by silencing all other audio tracks.

source Clips as they exist on their own and individually, not as part of a sequence.

storyboard A sequence of sketches that indicates the look and the progression of shots, used as a visual planning tool.

sweetening Improving the sound of audio in a sequence, such as adjusting levels, adjusting frequency equalization, or adding effects.

temperature In image editing, the warmth or coolness of the overall color balance in a frame, along a blue-yellow axis. Based on the concept of color temperature on the Kelvin scale.

thumbnail An icon that contains a small visual preview of a clip's contents.

TIFF (Tagged Image File Format) Graphics format for still images. Supports many types of compression, along with alpha channels for high-quality transparency. Suitable for print as well as video.

time remapping Speeding up or slowing down the playback speed of a clip over time, instead of changing playback to a single speed for its entire duration.

Timeline The part of the Timeline panel that contains clips sequenced along a time ruler, on one or more tracks. Timeline is sometimes used as a synonym for sequence.

tint The color balance in a frame along a magenta-green axis.

title Text superimposed on a video frame. Not only for the title of a video, it can refer to any text, including lower-thirds text and end credits, but not open or closed captions.

transcoding Re-encoding or converting clips to a different format. Often done for easier editing. For example, H.264 can be a processor-intensive codec for editing, so transcoding source clips to a Cineform or Apple ProRes codec can result in smoother editing.

transition A method of gradually transforming from one clip to another, added at an edit point. For example, during a crossfade transition, the first clip gradually disappears as the second clip gradually appears at the same time.

trim Adjust how much of a clip is used in a sequence by adjusting the In point or Out point.

Typekit An Adobe cloud service that provides access to a large font library. Premiere Pro and other Adobe software use Typekit to facilitate the replacement of missing fonts.

USB Universal Serial Bus, a protocol for connecting peripherals such as storage drives. For high-definition video editing, connecting storage drives using USB 3 or faster is preferred; USB 2 and earlier are too slow.

voiceover An audio clip containing spoken narration that plays during the video in a sequence.

workspace An arrangement of panels within a Premiere Pro application window.

ZIP file A document package, also called an archive, that can store one or more files using lossless compression. Combining files into a single ZIP file can simplify uploading and downloading.

Index

' (apostrophe) key, 117
\ (backslash) key, 56, 115
, (comma) key, 50, 56
. (period) key, 50, 56
~ (tilde) key, 48, 92, 226
1:1 aspect ratio, 190
16:9 aspect ratio, 190
180-degree rule, 259

A

about this book, viii–ix, 4–5
ACA Objectives, ix
action, match on, 260
action scene editing project, 149–201
    adjustment layer, 180–182
    audio editing, 177–180
    challenge exercise, 201
    credits, 183–186
    exporting multiple sequences, 197–198
    finishing sequence edits, 176–177
    importing reliable files, 159–160
    inspecting clip properties, 163–164
    media acquisition/creation, 151–153
    multicam sequence edits, 172–176
    preproduction process, 149–151
    Project Manager settings, 199–200
    proxies, 193–197
    recording a voiceover, 187–190
    relinking offline media, 160–162
    rough cut creation, 164–167
    setting up, 158–159
    Timeline productivity tips, 182–183
    unused clip cleanup, 197
    vertical video editing, 168–172
actor/talent, 247
Add Keyframe button, 129
Add Marker button, 132, 183
Add Tracks option, 212
Additive Dissolve transition, 78
additive editing, 105, 116
adjustment layers, 180–182
Adobe Add-ons, xiii
Adobe After Effects, 271–272
Adobe Application Manager, xi
Adobe Audition, 270–271
Adobe Bridge, 272–273
Adobe Capture app, 24, 274
Adobe Certified Associate (ACA) credential, xiii
Adobe Certified Associate (ACA) Exam, viii, 5
Adobe Character Animator, 271
Adobe Creative Cloud. *See* Creative Cloud
Adobe Forums, xiii, 25
Adobe Illustrator, 274
Adobe Learn books, 4
Adobe Media Encoder. *See* Media Encoder
Adobe Photoshop. *See* Photoshop
Adobe Portfolio, 276
Adobe Prelude, 270
Adobe Premiere Clip, 274
Adobe Premiere Pro. *See* Premiere Pro
Adobe Spark Video, 275
Adobe Stock, 17, 152, 275–276
Adobe Team Projects, 276–277
Adobe Typekit fonts, xii, 135
After Effects program, 271–272
AI file format, 139
Alpha Channel setting, 210–211
Ambience option, 179
anchor points, 213
animated logo, 213–215
appearance preferences, 155
application window size, 33
arrow keys, 53
art director, 247
artist release, 14
aspect ratios
    pixel, 164
    proxy, 196
    video, 190

Assembly workspace, 101–102
attention, directing, 252–254
attributes, copying/pasting, 232
audience, target, 244
Audio Clips bin, 55
Audio Clips folder, 11, 178
audio editing, 66–70, 106–110
    audio type assigned for, 109
    configuring hardware for, 155
    dragging and dropping clips for, 55
    effects applied in, 106–107
    Essential Sound panel for, 108–109
    meters displayed for, 68
    noise reduction in, 108–109
    setting up for, 177–178
    sound design tips for, 110
    sweetening audio in, 178–180
    Timeline panel controls for, 66
    transitions applied in, 78–79
    visibility options for, 67
    volume adjustments in, 68–70
    workspace setup for, 106
audio effects, 106–107
audio formats, 164
audio hardware preferences, 155, 187–188
Audio Meters panel, 29, 68
audio recording, 187–190
    audio hardware settings, 187–188
    persistent audio, 261
    Timeline settings, 188–189
    voiceover recording steps, 189–190
audio tracks
    export options for, 84, 85
    improving visibility of, 67, 107
    volume adjustments for, 68–70
Audio workspace, 106, 108, 177
Audition, Adobe, 270–271
Auto Tone setting, 113
Automate to Sequence option, 48, 227, 229
AVCHD format, 94

B

background music, 178, 179–180
backgrounds
    importing graphics with transparent, 139
    keying out green screen, 205, 210–212
Backspace key, 37

Balanced Background Music preset, 179
balanced composition, 257
before-and-after comparison, 113
bins, 38
    creating new, 40, 47–48
    organizing media into, 40–41
    origin of term, 93
    preferences for, 154
    rough cuts created from, 54–55
bird's-eye shot, 254
bitmap graphics, 274
Bitrate Encoding option, 232–233
black bars, 169–172
Blacks control, 113
Bold Title template, 184
breaking the rules, 262–263
Bridge, Adobe, 272–273
Brightness preference, 155
B-roll footage, 121–122, 261
budgeting projects, 248
Button Editor, 174, 194

C

cameras
    cutting between, 174–176
    stabilizing shake from, 139–141
caption export options, 85
Capture Format option, 21
challenge projects
    composited video, 221
    mini-documentary, 147
    multi-camera sequence, 201
    promotional video, 87
    slide show video, 236
Change Sequence Settings option, 98–99
Character Animator program, 271
chroma key compositing, 205
client job requirements, 243–244
clip markers, 132
Clip Mismatch Warning alert, 98–99
Clip Speed/Duration dialog box, 123
clips
    cleaning up unused, 197
    color labels for, 120
    deleting from Timeline, 53
    dragging and dropping, 52, 55
    effects added to, 75–76

instances of, 119
logging and naming, 9
marking parts of, 115–116
merging video and audio, 142–143
naming/renaming, 120
playback speed of, 123–125
previewing the contents of, 41
properties of, 163–164
rearranging in a sequence, 65
removing gaps between, 117
sequences based on, 46
setting In and Out points for, 116
shortcuts for adding, 50–51
stabilizing shaky, 139–141
subtracting parts of, 114–118
time remapping of, 127–130
trimming, 41, 60–63
Close button, 30
Close Gap command, 176–177
Close Panel command, 30
cloud storage, 17–18
cloud-based services, 275–277
collaboration process, 3
color correction, 110–114
    expressive adjustments for, 113–114
    Fast Color Corrector for, 206
    Lumetri Color panel for, 110–114
    manipulating color through, 262
    previewing and resetting changes for, 113
    tone and saturation options for, 112–113
    white balance fixes for, 111–112, 205–206
color grading, 262
color labels for clips, 120
Color workspace, 111
communication, effective, 248–249
compositing project, 203–221
    adding and animating graphics, 212–219
    challenge on creating your own, 221
    drawing a garbage matte, 207–209
    export process, 219–220
    green screen shooting guidelines, 207
    keying out the green background, 210–212
    preproduction process, 203–204
    setting up, 204–207
    See also green screen effects
composition principles, 256–259
context menus, 38
continuity editing, 260–261

Contrast control, 112
conventions used in book, ix
copying/pasting
    attributes, 232
    effects, 141, 232
copyrighted material, 264
Corner Pin effect, 172
costume designer, 247
Create Multi-Camera Source Sequence dialog box, 173
Create Proxies dialog box, 195
Creative adjustments, 113–114
Creative Cloud, 269–277
    cloud-based services, 275–277
    graphics applications, 273–274
    mobile apps, 274–275
    video production applications, 270–273
Creative Cloud (CC) Libraries, 24, 275
Creative Cloud desktop application, xi, 270
Creative Cloud Files storage, 17–18
Creative Commons licenses, 87, 264–265
credits, 183–186
    adding rolling, 185–186
    template for adding, 184–185
Cross Dissolve transition, 78
cross-cut, 261
customizing
    Project panel, 102
    Warp Stabilizer, 140–141
    workspaces, 99–101
cutaway shots, 261

D

deep focus shot, 252
Delete key, 37
deleting
    clips from Timeline panel, 53
    media cache files, 158
    ZIP files, 13
    See also removing
deliverables, 245
development phase, 239–240
Dialogue button, 109, 179
digital imaging technician (DIT), 246
directing attention, 252–254
director of photography (DP), 246–247
director role, 246

discontinuity, 262–263
distance, shot, 250–252
distractions, visual, 258–259
distribution phase, 242
Dockery, Joe, 87, 147, 154, 201, 221, 236
docking panels, 29, 30, 31–32
documentary project, 147
dolly shot, 253
Down Arrow key, 53
downloading lesson files, 10
dragging-and-dropping clips, 52, 55
drive system setup, 21–23
Drop Shadow effect, 218–219
Duplicate button, 220, 234
duration
    rolling credits, 186
    slide show image, 224
    title, 73
    transition, 77, 156
Dutch angle, 263

E

Ease In/Ease Out options, 186, 215, 231
eBook edition of book, viii, xi–xii
Edit Workspaces dialog box, 100–101
editing
    subtractive vs. additive, 105, 116
    titles, 72–73, 135–138
    See also audio editing; video editing
Editing workspace
    panel layout in, 35–36
    primary panels in, 26–28
editor, video, 247
educator resources, xiii
Effect Controls panel, 77, 79, 114, 171
effects
    adding to clips, 75–76
    adjustment layer, 181–182
    applying to exported videos, 84
    audio, 106–107
    copying/pasting, 141, 232
    Corner Pin, 172
    Drop Shadow, 218
    Gaussian Blur, 170
    Ken Burns, 230–232
    slide show motion, 230–232

    Ultra Key, 210–212
    Warp Stabilizer, 140–141
Effects panel, 28, 75–76, 107
Ellipse tool, 137
embedding content, 5–6
End key, 57
Esc key, 72
Essential Graphics panel, 72, 73, 134, 138, 184
Essential Sound panel
    overview on using, 108–109
    sweetening audio using, 178–180
establishing shot, 250
exclusive licenses, 265
Expand All Tracks option, 67
Export Frame button, 64, 131
Export Settings dialog box
    options available in, 83–86
    setting up an export in, 81–83, 197
exporting, 44
    Media Encoder for, 86, 143–146, 219–220, 232–235
    multiple sequences, 197–198, 232–235
    options available for, 83–86
    queueing media for, 144–145
    setting up sequences for, 80–83
    still frames, 131
    video slide shows, 232–235
Exports folder, 12–13
Exposure control, 112
Extract button, 117
eyedroppers
    Key Color, 210
    WB Selector, 112
eyeline match, 260

F

fair use, 264
Fast Color Corrector, 206
field of view, 250–252
file path, 163
files
    embedding vs. linking to, 5–6
    organizing into folders, 11–13
    selecting multiple, 12
    storage system for, 7–8
    See also media files

Fill Right with Left effect, 106–107
finding
    stock media files, 151–152
    templates, 17
flash forward, 261–262
flashback, 261
floating panels, 29, 30, 33
folders
    creating project, 9
    List view option for, 13
    organizing files into, 11–13
    See also bins
fonts, project, xii
forums, Adobe, xiii
Frame Blending option, 125
Frame Hold option, 131
frame rate, 163
Frame Sampling option, 125
frames
    exporting, 131
    extracting still, 64
    freezing, 130, 131
framing subjects, 256
Free Draw Bezier tool, 208–209
Freeze Frame icon, 130
freezing frames, 130, 131
fx badge, 128

G

gaffer, 247
gap removal, 118, 176–177
garbage matte, 207–209
Gaussian Blur effect, 170
General tab, 18–21
GIF file format, 139
GPU Acceleration option, 19–20, 114
graphics
    adding for green screen composite, 212–219
    Adobe applications for working with, 273–274
    filling vertical black bars with, 170–172
    importing with transparent backgrounds, 139
graphics card, 20
Graphics folder, 11
Graphics workspace, 134, 136, 184
gray target, 205

green screen effects
    adding and animating graphics for, 212–219
    drawing garbage mattes for, 207–209
    guidelines for shooting, 207
    keying out green backgrounds for, 210–212
    white balancing for, 205–206
    See also compositing project
grouping panels, 29, 30, 34

H

Hand tool, 56
hard disk drives (HDDs), 8
Help menu, 25
hidden tools, 58
high-angle shot, 254
Highlight color preference, 155
Highlights control, 113
Home key, 57, 134
hover-scrubbing clips, 41

I

Illustrator, Adobe, 274
Image Sequence option, 42
Import Layered File dialog box, 204–205
Import Presets button, 198
importing
    graphics with transparency, 139
    Media Browser options for, 91–93
    media files, 38–40, 42, 91–93, 159–160
In and Out points
    marking, 49–50, 115–116
    trimming clips using, 41
Incompetech website, 87, 152
Ingest option, Media Browser, 162
Ingest Settings tab, 24
Insert button, 50, 51
inspecting clip properties, 163–164
installing Premiere Pro, x
instances, clip, 119
integration process, 3
interview editing project, 89–147
    audio adjustments, 105–110
    clip playback speed, 123–125
    color correction, 110–114
    exporting with Media Encoder, 143–146
    L and J cuts, 120–122

markers, 131–133
Media Browser imports, 91–93
merging video and audio files, 142–143
mini-documentary challenge, 147
organization process, 119–120
playing a sequence smoothly, 125–127
preproduction process, 89–90
removing unwanted clip segments, 114–118
sequence creation, 95–99
setting up, 90–91
stabilizing shaky clips, 139–141
time remapping, 127–130
titles, 133–139
workspace, 99–104

J

J cuts, 120–122
JKL keyboard shortcuts, 55, 56
job requirements, 13–15, 243–245
    client, 243–244
    deliverable, 245
    purpose for video, 245
    target audience, 244
JPEG file format, 139
jump cuts, 118, 262

K

Ken Burns effect, 230–232
Key Color eyedropper, 210
keyboard shortcuts, 37
    for adding clips, 50
    for adding markers, 133
    for building rough cuts, 55–56
    for creating new files, 17
    for Media Browser imports, 93
    for opening files, 17
    for Timeline panel navigation, 56
    for zooming in/out, 116
keyframes
    audio, 68–69
    video, 128–129, 214
keying out backgrounds, 205
Keys to Success, 87, 147, 201, 221, 236

L

L cuts, 120–122
labeling clips, 120
layers
    applying adjustment, 180–182
    importing Photoshop documents as, 204–205
    list of Essential Graphics panel, 138
lead room, 258
leading lines, 258
Learn tab, Start screen, 17
legal issues, 14–15, 153, 263–266
    licenses, 14, 87, 264–265
    media attorney for, 265–266
    model and property releases, 14, 153, 265
lesson files, xi–xii, 10
licenses, 14, 87, 264–265
Lift button, 116
Link Media dialog box, 161
linked media files, 6, 160–162
Linked Selection button, 183
List view, 41–42, 47, 102–104
Locate File dialog box, 162
locating projects, 24–25
location manager, 247–248
Location option, 19
Lock icon, 65, 122
logo animation, 213–215
long shot, 250
low-angle shot, 255
lower-third titles, 133
Lumetri Color panel, 110–114

M

Mac computers, x, 36–38
    context menus, 38
    keyboard shortcuts, 37
    Preferences command, 38
managing media files, 5–9
Mark In/Out icons, 49–50, 115
markers, 131–133
    adding clip vs. sequence, 132
    editing or annotating, 133
    In and Out point, 49–50, 115–116
    sequence, 132–133, 228–229
    text notes added to, 230
match cut, 260

match on action, 260
Matte Cleanup options, 211
Media Browser panel, 42
    advantages of using, 94
    annotated illustration of, 92
    importing media with, 91–93
    Ingest option, 162
Media Cache Preferences panel, 156–158
Media Encoder, 272
    annotated illustration of, 145
    exporting with, 86, 143–146, 197–198, 219–220, 232–235
    multiple versions export using, 232–235
media files
    bins for organizing, 40–41
    creating your own, 153
    embedding vs. linking to, 5–6
    finding on stock websites, 151–152
    importing, 38–40, 42, 91–93, 159–160
    listing available/acquired, 14
    organizing into folders, 11–13
    preventing legal issues with, 14–15
    relinking offline, 160–162
    selecting multiple, 12
    storage system for, 7–8
    viewing imported, 41–42
Media Offline screen, 160
Media Preferences pane, 155–156
mediafiles folder, 11, 12
memorial slide show. *See* slide show video project
menus
    context, 38
    Help, 25
    overflow, 30
    Panel, 30
    Window, 26
    Workspaces, 35
Mercury Playback Engine, 20
Merge Clips dialog box, 142–143
metadata
    entering for video export, 82, 83
    List view display of, 104
Metadata Display dialog box, 104
Metadata Export dialog box, 82–83
mid-shot, 251
mini-documentary challenge, 147
mobile apps, 274–275

model release, 14, 153, 265
modifier keys, 37
monitors, multiple, 34
motion effects, slide show, 230–232
motion graphics template, 138–139
MPEG multiplexing options, 84, 85
multicam sequences, 172–176
    adding to the Timeline, 174
    creation of, 173, 201
    cutting between cameras, 174–176
    practicing edits with, 176
multiple-monitor setup, 34
music
    background, 178, 179–180
    Creative Commons licenses for, 87
    licensing concerns related to, 14
    video slide show, 226
Mute track button, 66

N

naming/renaming
    clips, 9, 120
    sequences, 47, 97
navigation
    techniques for time, 52
    Timeline panel, 56–57
nesting sequences, 190–192
    one inside another, 191–192
    social media adaptations, 191
network storage, 8
New Bin button, 40
New Project dialog box, 17, 18–24
    General tab, 18–21
    Ingest Settings tab, 24
    Scratch Disks tab, 21–23
New Sequence dialog box, 95–97, 225
noise reduction, 108–109
nonexclusive licenses, 265

O

one-drive system, 21
online storage, 17–18
opacity adjustments, 121, 213
opacity mask, 207–209
Open Project button, 17
operating system differences, x

Optical Flow option, 125
organizing
    clips in Timeline panel, 119–120
    files into folders, 11–13
    media into bins, 40–41
    project drives with teams, 8
overflow menu, 30
overscan effect, 74
over-the-shoulder shot, 255, 256
Overwrite button, 50

P

pan shot, 253
Panel menu, 30
panels, 26–34
    arranging, 29–34
    Audio Meters, 29, 68
    docking, 29, 30, 31–32
    Effect Controls, 77
    Effects, 28, 75–76
    Essential Graphics, 72, 73
    Essential Sound, 108–109, 178–180
    floating, 29, 30, 33
    grouping, 29, 30, 34
    layout of, 35–36
    Lumetri Color, 110–114
    maximizing, 48
    Media Browser, 42
    Preset Browser, 198
    Program, 26–27, 44
    Project, 28, 43
    Properties, 163–164
    Source, 26–27, 43
    Timeline, 26–27, 43–44
    Tools, 29, 58
    Workspaces, 35
panning, 230
Paste Attributes dialog box, 232
Pen tool, 68, 69, 137, 209
persistent audio, 261
Photoshop
    editing photographs with, 273
    importing layered documents from, 204–205
Pixabay website, 152
pixel aspect ratio, 164

playback
    smooth sequence, 125–127
    speed adjustments, 123–125
    time remapping, 127–130
playhead
    time indicator, 52, 57, 182
    tips for moving, 116
PNG file format, 139
points of view, 254–256
postproduction phase, 242
preferences, 154–158
    appearance, 155
    audio hardware, 155, 187–188
    general behavior, 154
    Media Cache, 156–158
    media scaling, 155–156
    opening on PC vs. Mac, 38
    resetting to default, xiii, 15
    Timeline panel, 156
Preferences command, 38
Prelude Live Logger, 274
Prelude, Adobe, 270
Premiere Pro
    installing, x
    operating system differences, x
    resetting preferences, xiii, 15
    resources, xii–xiii, 25, 277–278
    starting, 15–18
    updating, xi
    Windows vs. Mac, 36–38
pre-production folder, 11, 12
preproduction phase, 240–241
    action scene editing project, 149–151
    compositing project, 203–204
    interview editing project, 89–90
    slide show video project, 223
Preset Browser panel, 198
presets
    audio, 179
    changed settings in, 233
    export, 232–233, 235
    sequence, 96, 220, 225–226
previewing
    color corrections, 113
    transitions, 127
    video clips, 41
producer role, 246

production assistant, 247
production designer, 247
production phases, 239–242
    development, 239–240
    distribution, 242
    postproduction, 242
    preproduction, 240–241
    production, 241–242
production team, 246–248
Program Monitor controls, 53
Program panel, 26–27
    marking In and Out points in, 115
    video editing workflow and, 44
Project Auto Save option, 23
project files
    organizing, 90
    unzipping, 150
Project folder, 11
Project Manager, 199–200
Project panel, 28, 101–104
    customizing, 102
    importing media into, 38–40
    List view in, 41–42, 47, 102–104
    moving up a level in, 45
    rough cuts created from, 54–55
    video editing workflow and, 43
    working with media in, 41
Project Template folder, 12
projects
    creating new, 17
    editing settings for, 25
    folder management, 9
    locating, 24–25
    opening, 17
    sequences vs., 44
    setting up, 18–24
promotional video project, 87
Properties panel, 163–164
property release, 14, 265
proxies, 127, 193–197
    creating, 195–196
    setting up for, 194–195
    using in projects, 196–197
proxy workflow, 24, 193–194
PSD files, 139, 205
public domain media, 264
Publish video options, 85

R

rack focus shot, 253
Rate Stretch tool, 124, 208, 213
Razor tool, 63–64, 116
Recent projects list, 17
recording audio. *See* audio recording
Rectangle tool, 137
Reduce Noise option, 109
Reduce Rumble option, 109
relinking offline media, 160–162
Remove Unused Clips option, 197
removing
    gaps between clip sequences, 118
    unused clips from projects, 197
    unwanted clip segments, 114–118
    See also deleting
renaming. *See* naming/renaming
render bars, 126, 127
Renderer option, 19
Replace with Clip > From Bin command, 227, 228
Reset to Saved Layout option, 36
resetting preferences, xiii, 15
resizing application window, 33
resolution, playback, 126
Resolve Fonts dialog box, 135, 184
resources, xii–xiii, 25, 277–278
Reveal in Project option, 190
reverse-angle shot, 255
Ripple Delete option, 118
Ripple Edit tool, 61–62
ripple edits, 61–62
rolling credits, 185–186
Rolling Edit tool, 62–63
rolling edits, 62–63
rotation angles, 215
Rotation setting, 214
rough cuts, 49–56, 164–167
    building, 49–53, 166–167
    creating from Project panel, 54–55
    dragging and dropping clips into, 52
    keyboard shortcuts for, 55–56
    sequence creation for, 166
    storyboard for guiding, 165
royalty-free media, 151
rubber band control, 68–69
rule of thirds, 258

S

safe margins, 74–75, 134
saturation adjustments, 113
Save Preset button, 86
saving
    presets, 86
    workspaces, 35
scaling media, 156
scheduling projects, 241, 248
scratch disk setup, 21–23
    connections for, 23
    dual-drive system, 21–22
    single-drive system, 21
    three or more drive system, 22–23
Scratch Disks tab, 21–23
screenwriter, 246
script writing, 240
Selection tool, 59–61
    rearranging clips with, 65
    resizing titles with, 72
    trimming clips with, 60–61
sequence icon, 45, 47
sequence markers, 132
Sequence Settings dialog box, 46
sequences, 44–56
    adapting for social media, 191
    adding clips to, 50–51
    Automate to Sequence option, 48, 227, 229
    building rough cuts of, 49–56, 166–167
    Clip Mismatch Warning alert, 98–99
    creating new, 45–48, 95–98, 191
    dragging clips into, 52
    explanation of using, 44–45
    exporting, 80–83, 197–198
    interview project, 95–99
    markers in, 131–133, 228–229
    mixing media types in, 46
    naming/renaming, 47, 97
    nesting, 190–192
    playing smoothly, 125–127
    presets for, 96, 220, 225–226
    projects distinguished from, 44
    rearranging clips in, 65
    removing marked segments from, 116–118
    setting In and Out points for, 116
    slide show video, 225–229
    transitions added to, 76–77
    verifying settings for, 48
    vertical video added to, 168–169
    video clips for creating, 46
Shadows control, 113
shaky clip stabilization, 139–141
shallow focus shot, 252
shooting video, 250–256
    directing attention in, 252–254
    distance and field of view in, 250–252
    points of view in, 254–256
shot lists, 150–151
Show Video Keyframes option, 128
slide show video project, 223–236
    challenge on creating, 236
    default transition for, 229–230
    exporting multiple versions of, 232–235
    Ken Burns motion effect, 230–232
    preproduction process, 223
    sequence creation, 225–229
    setting up, 224
Slide tool, 64
Slip tool, 64
slow-motion clips, 123–125
Snap button, 183
social media uploads, 85, 191
solid-state drives (SSDs), 8, 23
Solo track button, 66
Sort button, Project panel, 102
Source Monitor
    adding clips to Timeline from, 51
    marking In and Out points in, 49–50
    overview of controls in, 53
Source panel, 26–27, 43
Spark Video app, 275
Spatial Interpolation commands, 231
speed adjustments
    clip playback, 123–124
    time remapping, 127–130
stabilizing shaky clips, 139–141
Start screen, 16–17
starting Premiere Pro, 15–18
startup preferences, 154
still frames
    duration setting, 156
    exporting, 131
    extracting, 64
stock media websites, 151–152

storing your files
    cloud storage for, 18
    multiple drives for, 7–8
Storyblocks website, 152
storyboards, 151, 165, 240
subtractive editing, 105, 116
sweetening audio, 178–180
symmetrical composition, 257
Sync Settings button, 18

T

talent/actor, 247
target audience, 244
Team Projects option, 18, 276–277
templates
    credits, 184–185
    export, 82
    finding, 17
    motion graphics, 138–139
Temporal Interpolation commands, 231
text
    adding, 70–72
    editing, 72–73
    See also credits; titles
Text tool, 136–137
three-point editing, 167
Thumbs.db files, 11
Thunderbolt connections, 23
TIFF file format, 139
tight shots, 250–251
time
    manipulation techniques, 261–262
    navigation techniques, 52
    notation conventions, 52
Time Remapping feature, 127–130
    freezing a frame, 130
    varying clip speed, 128–130
Timeline panel, 26–27
    adjusting options in, 156
    audio controls in, 66, 67, 188–189
    clips added to, 51
    display settings, 183
    multicam sequences in, 174
    navigation of, 56–57
    organizing clips in, 119–120
    overview of controls in, 57

productivity tips using, 182–183
rearranging clips in, 65
sequence creation in, 46
titles edited in, 72–73
trimming clips in, 60–61
video editing workflow and, 43–44
title bar, 24–25
titles, 133–139
    adding, 70–72, 134–135
    editing, 72–73, 135–138
    safe margins for, 74–75, 134
    saving as motion graphics template, 138–139
Toggle Multi-Camera View button, 175, 176
Toggle Proxies button, 194, 196
Tone settings, 112–113
tools, 57–65
    Ellipse, 137
    Free Draw Bezier, 208–209
    Hand, 56
    Pen, 68, 69, 137
    Rate Stretch, 124, 213
    Razor, 63–64
    Rectangle, 137
    Ripple Edit, 61–62
    Rolling Edit, 62–63
    Selection, 59–61
    Slide, 64
    Slip, 64
    Text, 136–137
    Track Select Forward, 65
    Type, 71, 185
    viewing hidden, 58
    Zoom, 56
Tools panel, 29, 58
Track Select Forward tool, 65
tracks. *See* audio tracks; video tracks
transcoded clips, 24
transitions, 76–80
    adding to sequences, 76–77
    applying audio, 78–79
    duration of, 77, 156
    extra time required for, 80
    previewing, 127
    setting up default, 78
    slide show video, 229–230
    wise use of, 79

trimming clips
   In and Out points for, 41
   Ripple Edit tool for, 61–62
   Rolling Edit tool for, 62–63
   Selection tool for, 60–61
Triple Scoop Music website, 152
tutorials, Premiere Pro, xii–xiii, 17
two-drive system, 21–22
two-pass encoding, 232
two-shot, 251
Type tool, 71, 185
Typekit fonts, xii, 135

U

Ultra Key effect, 210–212
unbalanced compositions, 263
undocking panels, 33
unit production manager, 248
unpacking ZIP files, 10–11
Unsplash website, 152
Up Arrow key, 53
updates, software, xi
USB connections, 23

V

variable frame rate, 164
vector graphics, 274
vertical video, 168–172
   adding to a sequence, 168–169
   filling the black bars in, 169–172
Video Clips bin, 47–48
Video Clips folder, 11
video editing
   exploring tools for, 57–65
   panel arrangement for, 26–27
   techniques used in, 260–263
   vertical video and, 168–172
   workflow for, 27–28, 43–44
   See also action scene editing project; interview editing project
video export options, 84
video industry, 239–266
   effective communication in, 248–249
   licensing, rights, and releases in, 263–266
   phases of production in, 239–242
   project job requirements in, 243–245
   roles of production team in, 246–248
   visual standards/techniques in, 250–263
   ways of working in, 266
video production applications, 270–273
video production team, 246–248
video sequences. *See* sequences
video slide show. *See* slide show video project
video tracks, 52, 70, 106, 122, 212
vignetting technique, 114
vision of productions, 249
visual standards/techniques, 250–263
   breaking the rules of, 262–263
   composition principles, 256–259
   editing techniques, 260–263
   types of shots, 250–256
voiceover audio, 187–190
   hardware settings for, 187–188
   steps for recording, 189–190
   Timeline settings for, 188–189
Voice-Over Record button, 66, 188–189, 190
volume adjustments
   track volume, 68–70
   voiceover audio, 190

W

Warp Stabilizer, 140–141
WB Selector eyedropper, 112
weather report video. *See* compositing project
Web Edition of book, viii, xi–xii
Welcome screen, 15
white balance adjustments
   color corrections and, 111–112
   gray target used for, 205–206
Whites control, 113
wide shot, 250, 251
Window menu, 26
Windows computers, x, 36–38
   context menus, 38
   keyboard shortcuts, 37
   Preferences command, 38
Work tab, Start screen, 17
workflow
   proxy, 24, 193–194
   subtractive, 105
   video editing, 27–28, 43–44

workspaces, 35–36
    Assembly, 101–102
    Audio, 106, 108, 177
    Color, 111
    customization of, 99–101
    Editing, 26–28, 35
    Graphics, 134, 136, 184
Workspaces bar, 100
Workspaces menu, 35
Workspaces panel, 35, 100
worm's-eye shot, 255
wrench icon, 67

Y
YouTube exports, 143–146, 235

Z
ZIP files
    archived project, 200
    deleting, 13
    unpacking, 10–11
zoom shot, 251–252
Zoom tool, 56
zooming in/out, 116, 230–231

v